T0192832

RILEM State-of-the-Art Reports

RILEM STATE-OF-THE-ART REPORTS
Volume 37

RILEM, The International Union of Laboratories and Experts in Construction Materials, Systems and Structures, founded in 1947, is a non-governmental scientific association whose goal is to contribute to progress in the construction sciences, techniques and industries, essentially by means of the communication it fosters between research and practice. RILEM's focus is on construction materials and their use in building and civil engineering structures, covering all phases of the building process from manufacture to use and recycling of materials. More information on RILEM and its previous publications can be found on www.RILEM.net.

The RILEM State-of-the-Art Reports (STAR) are produced by the Technical Committees. They represent one of the most important outputs that RILEM generates – high level scientific and engineering reports that provide cutting edge knowledge in a given field. The work of the TCs is one of RILEM's key functions.

Members of a TC are experts in their field and give their time freely to share their expertise. As a result, the broader scientific community benefits greatly from RILEM's activities.

RILEM's stated objective is to disseminate this information as widely as possible to the scientific community. RILEM therefore considers the STAR reports of its TCs as of highest importance, and encourages their publication whenever possible.

The information in this and similar reports is mostly pre-normative in the sense that it provides the underlying scientific fundamentals on which standards and codes of practice are based. Without such a solid scientific basis, construction practice will be less than efficient or economical.

It is RILEM's hope that this information will be of wide use to the scientific community.

Indexed in SCOPUS, Google Scholar and SpringerLink.

More information about this series at https://link.springer.com/bookseries/8780

Eddie Koenders · Kei-ichi Imamoto ·
Anthony Soive
Editors

Benchmarking Chloride Ingress Models on Real-life Case Studies—Marine Submerged and Road Sprayed Concrete Structures

State-of-the-Art Report of the RILEM TC 270-CIM

Editors
Eddie Koenders ⓘ
Institut Werkstoffe im Bauwesen
Technische Universitat Darmstadt
Darmstadt, Germany

Kei-ichi Imamoto ⓘ
Department of Architecture
Tokyo University of Science
Tokyo, Japan

Anthony Soive
Pôle d'activités Les Milles
Cerema Méditerranée
Aix-en-Provence, France

ISSN 2213-204X ISSN 2213-2031 (electronic)
RILEM State-of-the-Art Reports
ISBN 978-3-030-96424-5 ISBN 978-3-030-96422-1 (eBook)
https://doi.org/10.1007/978-3-030-96422-1

This Springer imprint is published by the registered company Springer Nature Switzerland AG
The registered company address is: Gewerbestrasse 11, 6330 Cham, Switzerland

TC 270-CIM Members

Dr. Anthony Soive, e-mail: anthony.soive@cerema.fr
Dr. Andre Monteiro, e-mail: avmonteiro@lnec.pt
Dr. Bahman Ghiassi, e-mail: Bahman.Ghiassi@nottingham.ac.uk
Prof. Carmen Andrade, e-mail: candrade@cimne.upc.edu
Prof. Christoph Gehlen, e-mail: gehlen@tum.de
Dr. Flavio Silva, e-mail: fsilva@puc-rio.br
Dr. Hans Beushausen, e-mail: hans.beushausen@uct.ac.za
Dr. Jonathan Mai-Nhu, e-mail: j.mai-nhu@cerib.com
Dr. Jose Pacheco, e-mail: jpacheco@ctlgroup.com
Dr. Ken-ichiro Nakarai, e-mail: nakarai@hiroshima-u.ac.jp
Dr. Sree Nanukuttan, e-mail: s.nanukuttan@qub.ac.uk
Prof. Eddie Koenders, e-mail: koenders@wib.tu-darmstadt.de
Dr. Henrik Erndahl Sørensen, e-mail: hks@teknologisk.dk
Dr. Jan Bisschop, e-mail: Jan.Bisschop@tfb.ch
Prof. Jason Weiss, e-mail: Jason.Weiss@oregonstate.edu
Dr. Jens Frederiksen, e-mail: jmfr@cowi.com
Dr. Joost Gulikers, e-mail: joost.gulikers@rws.nl
Prof. Kei-ichi Imamoto, e-mail: imamoto@rs.kagu.tus.ac.jp
Prof. Koichi Matsuzawa, e-mail: matmuner@gmail.com
Prof. Kyle Riding, e-mail: kyle.riding@essie.ufl.edu
Prof. Lars-Olof Nilsson, e-mail: lars-olof.nilsson@byggtek.lth.se
Dr. Miguel Azenha, e-mail: miguel.azenha@civil.uminho.pt
Dr. Philip van den Heede, e-mail: Philip.VandenHeede@UGent.be
Prof. Kefei LI, e-mail: likefei@tsinghua.edu.cn
Prof. M. Basheer, e-mail: P.A.M.Basheer@leeds.ac.uk
Prof. Takafumi Noguchi, e-mail: noguchi@bme.arch.t.u-tokyo.ac.jp
Prof. Toshiharu Kishi, e-mail: kishi@iis.u-tokyo.ac.jp
Dr. Qianqian Yu, e-mail: qianqian.yu@tongji.edu.cn
Prof. Qing-feng Liu, e-mail: liuqf@sjtu.edu.cn
Dr. Radhakrishna Pillai, e-mail: pillai@iitm.ac.in
Dr. Ravi Patel, e-mail: ravi.patel@kit.edu

Dr. Roberto Torrent, e-mail: torrent.concrete@gmail.com
Prof. Romildo Toledo Filho, e-mail: rdtoledofilho@gmail.com
Dr. Rui Miguel Lage Ferreira, e-mail: ruif@civil.ist.utl.pt
Prof. Tang Luping, e-mail: tang.luping@chalmers.se
Prof. Tetsuya Ishida, e-mail: tetsuya.ishida@civil.t.u-tokyo.ac.jp
Prof. Ueli Angst, e-mail: ueli.angst@ifb.baug.ethz.ch
Dr. Neven Ukrainczyk, e-mail: ukrainczyk@wib.tu-darmstadt.de
Dr. Zhang, e-mail: zhangys279@163.com

RILEM TC 270-CIM

Report: STAR
Date: 04.01.2022
Researchers who contributed actively to this STAR with model predictions:
Anass El Farissi, LASIE, France
André Monteiro, National Laboratory for Civil Engineering (LNEC), Portugal
Anthony Soive, Cerema, Aix-en-Provence, France
Bahman Ghiassi, University of Nottingham, UK
Jens Mejer Frederiksen, COWI, Lyngby, Denmark
Jonathan Mai-Nhu, Cerib, France
Lars-Olof Nilsson, Lund University, Sweden
Luping Tang, Chalmers University of Technology, Sweden
Neven Ukrainczyk, Technical University of Darmstadt, Germany
Qing-feng Liu, Shanghai Jiao Tong University, Shanghai, China
Quoc Huy Vu, Holcim, France
Ravi Patel, Karlsruhe Institute of Technology, Germany
Sreejith Nanukuttan, Queen's University Belfast, Ireland

RILEM Publications

The following list is presenting the global offer of RILEM Publications, sorted by series. Each publication is available in printed version and/or in online version.

RILEM Proceedings (PRO)

PRO 1: Durability of High Performance Concrete (ISBN: 2-912143-03-9; e-ISBN: 2-351580-12-5; e-ISBN: 2351580125); *Ed. H. Sommer*

PRO 2: Chloride Penetration into Concrete (ISBN: 2-912143-00-04; e-ISBN: 2912143454); *Eds. L.-O. Nilsson and J.-P. Ollivier*

PRO 3: Evaluation and Strengthening of Existing Masonry Structures (ISBN: 2-912143-02-0; e-ISBN: 2351580141); *Eds. L. Binda and C. Modena*

PRO 4: Concrete: From Material to Structure (ISBN: 2-912143-04-7; e-ISBN: 2351580206); *Eds. J.-P. Bournazel and Y. Malier*

PRO 5: The Role of Admixtures in High Performance Concrete (ISBN: 2-912143-05-5; e-ISBN: 2351580214); *Eds. J. G. Cabrera and R. Rivera-Villarreal*

PRO 6: High Performance Fiber Reinforced Cement Composites—HPFRCC 3 (ISBN: 2-912143-06-3; e-ISBN: 2351580222); *Eds. H. W. Reinhardt and A. E. Naaman*

PRO 7: 1st International RILEM Symposium on Self-Compacting Concrete (ISBN: 2-912143-09-8; e-ISBN: 2912143721); *Eds. Å. Skarendahl and Ö. Petersson*

PRO 8: International RILEM Symposium on Timber Engineering (ISBN: 2-912143-10-1; e-ISBN: 2351580230); *Ed. L. Boström*

PRO 9: 2nd International RILEM Symposium on Adhesion between Polymers and Concrete ISAP '99 (ISBN: 2-912143-11-X; e-ISBN: 2351580249); *Eds. Y. Ohama and M. Puterman*

PRO 10: 3rd International RILEM Symposium on Durability of Building and Construction Sealants (ISBN: 2-912143-13-6; e-ISBN: 2351580257); *Ed. A. T. Wolf*

PRO 11: 4th International RILEM Conference on Reflective Cracking in Pavements (ISBN: 2-912143-14-4; e-ISBN: 2351580265); *Eds. A. O. Abd El Halim, D. A. Taylor and El H. H. Mohamed*

PRO 12: International RILEM Workshop on Historic Mortars: Characteristics and Tests (ISBN: 2-912143-15-2; e-ISBN: 2351580273); *Eds. P. Bartos, C. Groot and J. J. Hughes*

PRO 13: 2nd International RILEM Symposium on Hydration and Setting (ISBN: 2-912143-16-0; e-ISBN: 2351580281); *Ed. A. Nonat*

PRO 14: Integrated Life-Cycle Design of Materials and Structures—ILCDES 2000 (ISBN: 951-758-408-3; e-ISBN: 235158029X); (ISSN: 0356-9403); *Ed. S. Sarja*

PRO 15: Fifth RILEM Symposium on Fibre-Reinforced Concretes (FRC)—BEFIB'2000 (ISBN: 2-912143-18-7; e-ISBN: 291214373X); *Eds. P. Rossi and G. Chanvillard*

PRO 16: Life Prediction and Management of Concrete Structures (ISBN: 2-912143-19-5; e-ISBN: 2351580303); *Ed. D. Naus*

PRO 17: Shrinkage of Concrete—Shrinkage 2000 (ISBN: 2-912143-20-9; e-ISBN: 2351580311); *Eds. V. Baroghel-Bouny and P.-C. Aïtcin*

PRO 18: Measurement and Interpretation of the On-Site Corrosion Rate (ISBN: 2-912143-21-7; e-ISBN: 235158032X); *Eds. C. Andrade, C. Alonso, J. Fullea, J. Polimon and J. Rodriguez*

PRO 19: Testing and Modelling the Chloride Ingress into Concrete (ISBN: 2-912143-22-5; e-ISBN: 2351580338); *Eds. C. Andrade and J. Kropp*

PRO 20: 1st International RILEM Workshop on Microbial Impacts on Building Materials (CD 02) (e-ISBN 978-2-35158-013-4); *Ed. M. Ribas Silva*

PRO 21: International RILEM Symposium on Connections between Steel and Concrete (ISBN: 2-912143-25-X; e-ISBN: 2351580346); *Ed. R. Eligehausen*

PRO 22: International RILEM Symposium on Joints in Timber Structures (ISBN: 2-912143-28-4; e-ISBN: 2351580354); *Eds. S. Aicher and H.-W. Reinhardt*

PRO 23: International RILEM Conference on Early Age Cracking in Cementitious Systems (ISBN: 2-912143-29-2; e-ISBN: 2351580362); *Eds. K. Kovler and A. Bentur*

PRO 24: 2nd International RILEM Workshop on Frost Resistance of Concrete (ISBN: 2-912143-30-6; e-ISBN: 2351580370); *Eds. M. J. Setzer, R. Auberg and H.-J. Keck*

PRO 25: International RILEM Workshop on Frost Damage in Concrete (ISBN: 2-912143-31-4; e-ISBN: 2351580389); *Eds. D. J. Janssen, M. J. Setzer and M. B. Snyder*

PRO 26: International RILEM Workshop on On-Site Control and Evaluation of Masonry Structures (ISBN: 2-912143-34-9; e-ISBN: 2351580141); *Eds. L. Binda and R. C. de Vekey*

PRO 27: International RILEM Symposium on Building Joint Sealants (CD03; e-ISBN: 235158015X); *Ed. A. T. Wolf*

PRO 28: 6th International RILEM Symposium on Performance Testing and Evaluation of Bituminous Materials—PTEBM'03 (ISBN: 2-912143-35-7; e-ISBN: 978-2-912143-77-8); *Ed. M. N. Partl*

PRO 29: 2nd International RILEM Workshop on Life Prediction and Ageing Management of Concrete Structures (ISBN: 2-912143-36-5; e-ISBN: 2912143780); *Ed. D. J. Naus*

PRO 30: 4th International RILEM Workshop on High Performance Fiber Reinforced Cement Composites—HPFRCC 4 (ISBN: 2-912143-37-3; e-ISBN: 2912143799); *Eds. A. E. Naaman and H. W. Reinhardt*

PRO 31: International RILEM Workshop on Test and Design Methods for Steel Fibre Reinforced Concrete: Background and Experiences (ISBN: 2-912143-38-1; e-ISBN: 2351580168); *Eds. B. Schnütgen and L. Vandewalle*

PRO 32: International Conference on Advances in Concrete and Structures 2 vol. (ISBN (set): 2-912143-41-1; e-ISBN: 2351580176); *Eds. Ying-shu Yuan, Surendra P. Shah and Heng-lin Lü*

PRO 33: 3rd International Symposium on Self-Compacting Concrete (ISBN: 2-912143-42-X; e-ISBN: 2912143713); *Eds. Ó. Wallevik and I. Níelsson*

PRO 34: International RILEM Conference on Microbial Impact on Building Materials (ISBN: 2-912143-43-8; e-ISBN: 2351580184); *Ed. M. Ribas Silva*

PRO 35: International RILEM TC 186-ISA on Internal Sulfate Attack and Delayed Ettringite Formation (ISBN: 2-912143-44-6; e-ISBN: 2912143802); *Eds. K. Scrivener and J. Skalny*

PRO 36: International RILEM Symposium on Concrete Science and Engineering—A Tribute to Arnon Bentur (ISBN: 2-912143-46-2; e-ISBN: 2912143586); *Eds. K. Kovler, J. Marchand, S. Mindess and J. Weiss*

PRO 37: 5th International RILEM Conference on Cracking in Pavements—Mitigation, Risk Assessment and Prevention (ISBN: 2-912143-47-0; e-ISBN: 2912143764); *Eds. C. Petit, I. Al-Qadi and A. Millien*

PRO 38: 3rd International RILEM Workshop on Testing and Modelling the Chloride Ingress into Concrete (ISBN: 2-912143-48-9; e-ISBN: 2912143578); *Eds. C. Andrade and J. Kropp*

PRO 39: 6th International RILEM Symposium on Fibre-Reinforced Concretes—BEFIB 2004 (ISBN: 2-912143-51-9; e-ISBN: 2912143748); *Eds. M. Di Prisco, R. Felicetti and G. A. Plizzari*

PRO 40: International RILEM Conference on the Use of Recycled Materials in Buildings and Structures (ISBN: 2-912143-52-7; e-ISBN: 2912143756); *Eds. E. Vázquez, Ch. F. Hendriks and G. M. T. Janssen*

PRO 41: RILEM International Symposium on Environment-Conscious Materials and Systems for Sustainable Development (ISBN: 2-912143-55-1; e-ISBN: 2912143640); *Eds. N. Kashino and Y. Ohama*

PRO 42: SCC'2005—China: 1st International Symposium on Design, Performance and Use of Self-Consolidating Concrete (ISBN: 2-912143-61-6; e-ISBN: 2912143624); *Eds. Zhiwu Yu, Caijun Shi, Kamal Henri Khayat and Youjun Xie*

PRO 43: International RILEM Workshop on Bonded Concrete Overlays (e-ISBN: 2-912143-83-7); *Eds. J. L. Granju and J. Silfwerbrand*

PRO 44: 2nd International RILEM Workshop on Microbial Impacts on Building Materials (CD11) (e-ISBN: 2-912143-84-5); *Ed. M. Ribas Silva*

PRO 45: 2nd International Symposium on Nanotechnology in Construction, Bilbao (ISBN: 2-912143-87-X; e-ISBN: 2912143888); *Eds. Peter J. M. Bartos, Yolanda de Miguel and Antonio Porro*

PRO 46: Concrete Life'06—International RILEM-JCI Seminar on Concrete Durability and Service Life Planning: Curing, Crack Control, Performance in Harsh Environments (ISBN: 2-912143-89-6; e-ISBN: 291214390X); *Ed. K. Kovler*

PRO 47: International RILEM Workshop on Performance Based Evaluation and Indicators for Concrete Durability (ISBN: 978-2-912143-95-2; e-ISBN: 9782912143969); *Eds. V. Baroghel-Bouny, C. Andrade, R. Torrent and K. Scrivener*

PRO 48: 1st International RILEM Symposium on Advances in Concrete through Science and Engineering (e-ISBN: 2-912143-92-6); *Eds. J. Weiss, K. Kovler, J. Marchand, and S. Mindess*

PRO 49: International RILEM Workshop on High Performance Fiber Reinforced Cementitious Composites in Structural Applications (ISBN: 2-912143-93-4; e-ISBN: 2912143942); *Eds. G. Fischer and V. C. Li*

PRO 50: 1st International RILEM Symposium on Textile Reinforced Concrete (ISBN: 2-912143-97-7; e-ISBN: 2351580087); *Eds. Josef Hegger, Wolfgang Brameshuber and Norbert Will*

PRO 51: 2nd International Symposium on Advances in Concrete through Science and Engineering (ISBN: 2-35158-003-6; e-ISBN: 2-35158-002-8); *Eds. J. Marchand, B. Bissonnette, R. Gagné, M. Jolin and F. Paradis*

PRO 52: Volume Changes of Hardening Concrete: Testing and Mitigation (ISBN: 2-35158-004-4; e-ISBN: 2-35158-005-2); *Eds. O. M. Jensen, P. Lura and K. Kovler*

PRO 53: High Performance Fiber Reinforced Cement Composites—HPFRCC5 (ISBN: 978-2-35158-046-2; e-ISBN: 978-2-35158-089-9); *Eds. H. W. Reinhardt and A. E. Naaman*

PRO 54: 5th International RILEM Symposium on Self-Compacting Concrete (ISBN: 978-2-35158-047-9; e-ISBN: 978-2-35158-088-2); *Eds. G. De Schutter and V. Boel*

PRO 55: International RILEM Symposium Photocatalysis, Environment and Construction Materials (ISBN: 978-2-35158-056-1; e-ISBN: 978-2-35158-057-8); *Eds. P. Baglioni and L. Cassar*

PRO 56: International RILEM Workshop on Integral Service Life Modelling of Concrete Structures (ISBN 978-2-35158-058-5; e-ISBN: 978-2-35158-090-5); *Eds. R. M. Ferreira, J. Gulikers and C. Andrade*

PRO 57: RILEM Workshop on Performance of cement-based materials in aggressive aqueous environments (e-ISBN: 978-2-35158-059-2); *Ed. N. De Belie*

PRO 58: International RILEM Symposium on Concrete Modelling—CONMOD'08 (ISBN: 978-2-35158-060-8; e-ISBN: 978-2-35158-076-9); *Eds. E. Schlangen and G. De Schutter*

PRO 59: International RILEM Conference on On Site Assessment of Concrete, Masonry and Timber Structures—SACoMaTiS 2008 (ISBN set: 978-2-35158-061-5; e-ISBN: 978-2-35158-075-2); *Eds. L. Binda, M. di Prisco and R. Felicetti*

PRO 60: Seventh RILEM International Symposium on Fibre Reinforced Concrete: Design and Applications—BEFIB 2008 (ISBN: 978-2-35158-064-6; e-ISBN: 978-2-35158-086-8); *Ed. R. Gettu*

PRO 61: 1st International Conference on Microstructure Related Durability of Cementitious Composites 2 vol., (ISBN: 978-2-35158-065-3; e-ISBN: 978-2-35158-084-4); *Eds. W. Sun, K. van Breugel, C. Miao, G. Ye and H. Chen*

PRO 62: NSF/ RILEM Workshop: In-situ Evaluation of Historic Wood and Masonry Structures (e-ISBN: 978-2-35158-068-4); *Eds. B. Kasal, R. Anthony and M. Drdácký*

PRO 63: Concrete in Aggressive Aqueous Environments: Performance, Testing and Modelling, 2 vol., (ISBN: 978-2-35158-071-4; e-ISBN: 978-2-35158-082-0); *Eds. M. G. Alexander and A. Bertron*

PRO 64: Long Term Performance of Cementitious Barriers and Reinforced Concrete in Nuclear Power Plants and Waste Management—NUCPERF 2009 (ISBN: 978-2-35158-072-1; e-ISBN: 978-2-35158-087-5); *Eds. V. L'Hostis, R. Gens and C. Gallé*

PRO 65: Design Performance and Use of Self-consolidating Concrete—SCC'2009 (ISBN: 978-2-35158-073-8; e-ISBN: 978-2-35158-093-6); *Eds. C. Shi, Z. Yu, K. H. Khayat and P. Yan*

PRO 66: 2nd International RILEM Workshop on Concrete Durability and Service Life Planning—ConcreteLife'09 (ISBN: 978-2-35158-074-5; ISBN: 978-2-35158-074-5); *Ed. K. Kovler*

PRO 67: Repairs Mortars for Historic Masonry (e-ISBN: 978-2-35158-083-7); *Ed. C. Groot*

PRO 68: Proceedings of the 3rd International RILEM Symposium on 'Rheology of Cement Suspensions such as Fresh Concrete (ISBN 978-2-35158-091-2; e-ISBN: 978-2-35158-092-9); *Eds. O. H. Wallevik, S. Kubens and S. Oesterheld*

PRO 69: 3rd International PhD Student Workshop on 'Modelling the Durability of Reinforced Concrete (ISBN: 978-2-35158-095-0); *Eds. R. M. Ferreira, J. Gulikers and C. Andrade*

PRO 70: 2nd International Conference on 'Service Life Design for Infrastructure' (ISBN set: 978-2-35158-096-7, e-ISBN: 978-2-35158-097-4); *Eds. K. van Breugel, G. Ye and Y. Yuan*

PRO 71: Advances in Civil Engineering Materials—The 50-year Teaching Anniversary of Prof. Sun Wei' (ISBN: 978-2-35158-098-1; e-ISBN: 978-2-35158-099-8); *Eds. C. Miao, G. Ye and H. Chen*

PRO 72: First International Conference on 'Advances in Chemically-Activated Materials—CAM'2010' (2010), 264 pp., ISBN: 978-2-35158-101-8; e-ISBN: 978-2-35158-115-5; *Eds. Caijun Shi and Xiaodong Shen*

PRO 73: 2nd International Conference on 'Waste Engineering and Management—ICWEM 2010' (2010), 894 pp., ISBN: 978-2-35158-102-5; e-ISBN: 978-2-35158-103-2, *Eds. J. Zh. Xiao, Y. Zhang, M. S. Cheung and R. Chu*

PRO 74: International RILEM Conference on 'Use of Superabsorbent Polymers and Other New Additives in Concrete' (2010) 374 pp., ISBN: 978-2-35158-104-9; e-ISBN: 978-2-35158-105-6; *Eds. O.M. Jensen, M.T. Hasholt, and S. Laustsen*

PRO 75: International Conference on 'Material Science—2nd ICTRC—Textile Reinforced Concrete—Theme 1' (2010) 436 pp., ISBN: 978-2-35158-106-3; e-ISBN: 978-2-35158-107-0; *Ed. W. Brameshuber*

PRO 76: International Conference on 'Material Science—HetMat—Modelling of Heterogeneous Materials—Theme 2' (2010) 255 pp., ISBN: 978-2-35158-108-7; e-ISBN: 978-2-35158-109-4; *Ed. W. Brameshuber*

PRO 77: International Conference on 'Material Science—AdIPoC—Additions Improving Properties of Concrete—Theme 3' (2010) 459 pp., ISBN: 978-2-35158-110-0; e-ISBN: 978-2-35158-111-7; *Ed. W. Brameshuber*

PRO 78: 2nd Historic Mortars Conference and RILEM TC 203-RHM Final Workshop—HMC2010 (2010) 1416 pp., e-ISBN: 978-2-35158-112-4; *Eds. J. Válek, C. Groot and J. J. Hughes*

PRO 79: International RILEM Conference on Advances in Construction Materials Through Science and Engineering (2011) 213 pp., ISBN: 978-2-35158-116-2, e-ISBN: 978-2-35158-117-9; *Eds. Christopher Leung and K.T. Wan*

PRO 80: 2nd International RILEM Conference on Concrete Spalling due to Fire Exposure (2011) 453 pp., ISBN: 978-2-35158-118-6; e-ISBN: 978-2-35158-119-3; *Eds. E.A.B. Koenders and F. Dehn*

PRO 81: 2nd International RILEM Conference on Strain Hardening Cementitious Composites (SHCC2-Rio) (2011) 451 pp., ISBN: 978-2-35158-120-9; e-ISBN: 978-2-35158-121-6; *Eds. R.D. Toledo Filho, F.A. Silva, E.A.B. Koenders and E.M.R. Fairbairn*

PRO 82: 2nd International RILEM Conference on Progress of Recycling in the Built Environment (2011) 507 pp., e-ISBN: 978-2-35158-122-3; *Eds. V.M. John, E. Vazquez, S.C. Angulo and C. Ulsen*

PRO 83: 2nd International Conference on Microstructural-related Durability of Cementitious Composites (2012) 250 pp., ISBN: 978-2-35158-129-2; e-ISBN: 978-2-35158-123-0; *Eds. G. Ye, K. van Breugel, W. Sun and C. Miao*

PRO 84: CONSEC13—Seventh International Conference on Concrete under Severe Conditions—Environment and Loading (2013) 1930 pp., ISBN: 978-2-35158-124-7; e-ISBN: 978-2- 35158-134-6; *Eds. Z.J. Li, W. Sun, C.W. Miao, K. Sakai, O.E. Gjorv and N. Banthia*

PRO 85: RILEM-JCI International Workshop on Crack Control of Mass Concrete and Related issues concerning Early-Age of Concrete Structures—ConCrack 3—Control of Cracking in Concrete Structures 3 (2012) 237 pp., ISBN: 978-2-35158-125-4; e-ISBN: 978-2-35158-126-1; *Eds. F. Toutlemonde and J.-M. Torrenti*

PRO 86: International Symposium on Life Cycle Assessment and Construction (2012) 414 pp., ISBN: 978-2-35158-127-8, e-ISBN: 978-2-35158-128-5; *Eds. A. Ventura and C. de la Roche*

PRO 87: UHPFRC 2013—RILEM-fib-AFGC International Symposium on Ultra-High Performance Fibre-Reinforced Concrete (2013), ISBN: 978-2-35158-130-8, e-ISBN: 978-2-35158-131-5; *Eds. F. Toutlemonde*

PRO 88: 8th RILEM International Symposium on Fibre Reinforced Concrete (2012) 344 pp., ISBN: 978-2-35158-132-2; e-ISBN: 978-2-35158-133-9; *Eds. Joaquim A.O. Barros*

PRO 89: RILEM International workshop on performance-based specification and control of concrete durability (2014) 678 pp., ISBN: 978-2-35158-135-3; e-ISBN: 978-2-35158-136-0; *Eds. D. Bjegović, H. Beushausen and M. Serdar*

PRO 90: 7th RILEM International Conference on Self-Compacting Concrete and of the 1st RILEM International Conference on Rheology and Processing of Construction Materials (2013) 396 pp., ISBN: 978-2-35158-137-7; e-ISBN: 978-2-35158-138-4; *Eds. Nicolas Roussel and Hela Bessaies-Bey*

PRO 91: CONMOD 2014—RILEM International Symposium on Concrete Modelling (2014), ISBN: 978-2-35158-139-1; e-ISBN: 978-2-35158-140-7; *Eds. Kefei Li, Peiyu Yan and Rongwei Yang*

PRO 92: CAM 2014—2nd International Conference on advances in chemically-activated materials (2014) 392 pp., ISBN: 978-2-35158-141-4; e-ISBN: 978-2-35158-142-1; *Eds. Caijun Shi and Xiadong Shen*

PRO 93: SCC 2014—3rd International Symposium on Design, Performance and Use of Self-Consolidating Concrete (2014) 438 pp., ISBN: 978-2-35158-143-8; e-ISBN: 978-2-35158-144-5; *Eds. Caijun Shi, Zhihua Ou and Kamal H. Khayat*

PRO 94 (online version): HPFRCC-7—7th RILEM conference on High performance fiber reinforced cement composites (2015), e-ISBN: 978-2-35158-146-9; *Eds. H.W. Reinhardt, G.J. Parra-Montesinos and H. Garrecht*

PRO 95: International RILEM Conference on Application of superabsorbent polymers and other new admixtures in concrete construction (2014), ISBN: 978-2-35158-147-6; e-ISBN: 978-2-35158-148-3; *Eds. Viktor Mechtcherine and Christof Schroefl*

PRO 96 (online version): XIII DBMC: XIII International Conference on Durability of Building Materials and Components (2015), e-ISBN: 978-2-35158-149-0; *Eds. M. Quattrone and V.M. John*

PRO 97: SHCC3—3rd International RILEM Conference on Strain Hardening Cementitious Composites (2014), ISBN: 978-2-35158-150-6; e-ISBN: 978-2-35158-151-3; *Eds. E. Schlangen, M.G. Sierra Beltran, M. Lukovic and G. Ye*

PRO 98: FERRO-11—11th International Symposium on Ferrocement and 3rd ICTRC—International Conference on Textile Reinforced Concrete (2015), ISBN: 978-2-35158-152-0; e-ISBN: 978-2-35158-153-7; *Ed. W. Brameshuber*

PRO 99 (online version): ICBBM 2015—1st International Conference on Bio-Based Building Materials (2015), e-ISBN: 978-2-35158-154-4; *Eds. S. Amziane and M. Sonebi*

PRO 100: SCC16—RILEM Self-Consolidating Concrete Conference (2016), ISBN: 978-2-35158-156-8; e-ISBN: 978-2-35158-157-5; *Ed. Kamal H. Kayat*

PRO 101 (online version): III Progress of Recycling in the Built Environment (2015), e-ISBN: 978-2-35158-158-2; *Eds I. Martins, C. Ulsen and S. C. Angulo*

PRO 102 (online version): RILEM Conference on Microorganisms-Cementitious Materials Interactions (2016), e-ISBN: 978-2-35158-160-5; *Eds. Alexandra Bertron, Henk Jonkers and Virginie Wiktor*

PRO 103 (online version): ACESC'16—Advances in Civil Engineering and Sustainable Construction (2016), e-ISBN: 978-2-35158-161-2; *Eds. T.Ch. Madhavi, G. Prabhakar, Santhosh Ram and P.M. Rameshwaran*

PRO 104 (online version): SSCS'2015—Numerical Modeling—Strategies for Sustainable Concrete Structures (2015), e-ISBN: 978-2-35158-162-9

PRO 105: 1st International Conference on UHPC Materials and Structures (2016), ISBN: 978-2-35158-164-3; e-ISBN: 978-2-35158-165-0

PRO 106: AFGC-ACI-fib-RILEM International Conference on Ultra-High-Performance Fibre-Reinforced Concrete—UHPFRC 2017 (2017), ISBN: 978-2-35158-166-7; e-ISBN: 978-2-35158-167-4; *Eds. François Toutlemonde and Jacques Resplendino*

PRO 107 (online version): XIV DBMC—14th International Conference on Durability of Building Materials and Components (2017), e-ISBN: 978-2-35158-159-9; *Eds. Geert De Schutter, Nele De Belie, Arnold Janssens and Nathan Van Den Bossche*

PRO 108: MSSCE 2016—Innovation of Teaching in Materials and Structures (2016), ISBN: 978-2-35158-178-0; e-ISBN: 978-2-35158-179-7; *Ed. Per Goltermann*

PRO 109 (2 volumes): MSSCE 2016—Service Life of Cement-Based Materials and Structures (2016), ISBN Vol. 1: 978-2-35158-170-4; Vol. 2: 978-2-35158-171-4; Set Vol. 1&2: 978-2-35158-172-8; e-ISBN : 978-2-35158-173-5; *Eds. Miguel Azenha, Ivan Gabrijel, Dirk Schlicke, Terje Kanstad and Ole Mejlhede Jensen*

PRO 110: MSSCE 2016—Historical Masonry (2016), ISBN: 978-2-35158-178-0; e-ISBN: 978-2-35158-179-7; *Eds. Inge Rörig-Dalgaard and Ioannis Ioannou*

PRO 111: MSSCE 2016—Electrochemistry in Civil Engineering (2016); ISBN: 978-2-35158-176-6; e-ISBN: 978-2-35158-177-3; *Ed. Lisbeth M. Ottosen*

PRO 112: MSSCE 2016—Moisture in Materials and Structures (2016), ISBN: 978-2-35158-178-0; e-ISBN: 978-2-35158-179-7; *Eds. Kurt Kielsgaard Hansen, Carsten Rode and Lars-Olof Nilsson*

PRO 113: MSSCE 2016—Concrete with Supplementary Cementitious Materials (2016), ISBN: 978-2-35158-178-0; e-ISBN: 978-2-35158-179-7; *Eds. Ole Mejlhede Jensen, Konstantin Kovler and Nele De Belie*

PRO 114: MSSCE 2016—Frost Action in Concrete (2016), ISBN: 978-2-35158-182-7; e-ISBN: 978-2-35158-183-4; *Eds. Marianne Tange Hasholt, Katja Fridh and R. Doug Hooton*

PRO 115: MSSCE 2016—Fresh Concrete (2016), ISBN: 978-2-35158-184-1; e-ISBN: 978-2-35158-185-8; *Eds. Lars N. Thrane, Claus Pade, Oldrich Svec and Nicolas Roussel*

PRO 116: BEFIB 2016—9th RILEM International Symposium on Fiber Reinforced Concrete (2016), ISBN: 978-2-35158-187-2; e-ISBN: 978-2-35158-186-5; *Eds. N. Banthia, M. di Prisco and S. Soleimani-Dashtaki*

PRO 117: 3rd International RILEM Conference on Microstructure Related Durability of Cementitious Composites (2016), ISBN: 978-2-35158-188-9; e-ISBN: 978-2-35158-189-6; *Eds. Changwen Miao, Wei Sun, Jiaping Liu, Huisu Chen, Guang Ye and Klaas van Breugel*

PRO 118 (4 volumes): International Conference on Advances in Construction Materials and Systems (2017), ISBN Set: 978-2-35158-190-2; Vol. 1: 978-2-35158-193-3; Vol. 2: 978-2-35158-194-0; Vol. 3: ISBN:978-2-35158-195-7; Vol. 4: ISBN:978-2-35158-196-4; e-ISBN: 978-2-35158-191-9; *Ed. Manu Santhanam*

PRO 119 (online version): ICBBM 2017—Second International RILEM Conference on Bio-based Building Materials, (2017), e-ISBN: 978-2-35158-192-6; *Ed. Sofiane Amziane*

PRO 120 (2 volumes): EAC-02—2nd International RILEM/COST Conference on Early Age Cracking and Serviceability in Cement-based Materials and Structures, (2017), Vol. 1: 978-2-35158-199-5, Vol. 2: 978-2-35158-200-8, Set: 978-2-35158-197-1, e-ISBN: 978-2-35158-198-8; *Eds. Stéphanie Staquet and Dimitrios Aggelis*

PRO 121 (2 volumes): SynerCrete18: Interdisciplinary Approaches for Cement-based Materials and Structural Concrete: Synergizing Expertise and Bridging Scales of Space and Time, (2018), Set: 978-2-35158-202-2, Vol.1: 978-2-35158-211-4, Vol.2: 978-2-35158-212-1, e-ISBN: 978-2-35158-203-9; *Eds. Miguel Azenha, Dirk Schlicke, Farid Benboudjema, Agnieszka Knoppik*

PRO 122: SCC'2018 China—Fourth International Symposium on Design, Performance and Use of Self-Consolidating Concrete, (2018), ISBN: 978-2-35158-204-6, e-ISBN: 978-2-35158-205-3; *Eds. C. Shi, Z. Zhang, K. H. Khayat*

PRO 123: Final Conference of RILEM TC 253-MCI: Microorganisms-Cementitious Materials Interactions (2018), Set: 978-2-35158-207-7, Vol.1: 978-2-35158-209-1, Vol.2: 978-2-35158-210-7, e-ISBN: 978-2-35158-206-0; *Ed. Alexandra Bertron*

PRO 124 (online version): Fourth International Conference Progress of Recycling in the Built Environment (2018), e-ISBN: 978-2-35158-208-4; *Eds. Isabel M. Martins, Carina Ulsen, Yury Villagran*

PRO 125 (online version): SLD4—4th International Conference on Service Life Design for Infrastructures (2018), e-ISBN: 978-2-35158-213-8; *Eds. Guang Ye, Yong Yuan, Claudia Romero Rodriguez, Hongzhi Zhang, Branko Savija*

PRO 126: Workshop on Concrete Modelling and Material Behaviour in honor of Professor Klaas van Breugel (2018), ISBN: 978-2-35158-214-5, e-ISBN: 978-2-35158-215-2; *Ed. Guang Ye*

PRO 127 (online version): CONMOD2018—Symposium on Concrete Modelling (2018), e-ISBN: 978-2-35158-216-9; *Eds. Erik Schlangen, Geert de Schutter, Branko Savija, Hongzhi Zhang, Claudia Romero Rodriguez*

PRO 128: SMSS2019—International Conference on Sustainable Materials, Systems and Structures (2019), ISBN: 978-2-35158-217-6, e-ISBN: 978-2-35158-218-3

PRO 129: 2nd International Conference on UHPC Materials and Structures (UHPC2018-China), ISBN: 978-2-35158-219-0, e-ISBN: 978-2-35158-220-6

PRO 130: 5th Historic Mortars Conference (2019), ISBN: 978-2-35158-221-3, e-ISBN: 978-2-35158-222-0; *Eds. José Ignacio Álvarez, José María Fernández, Íñigo Navarro, Adrián Durán, Rafael Sirera*

PRO 131 (online version): 3rd International Conference on Bio-Based Building Materials (ICBBM2019), e-ISBN: 978-2-35158-229-9; *Eds. Mohammed Sonebi, Sofiane Amziane, Jonathan Page*

PRO 132: IRWRMC'18—International RILEM Workshop on Rheological Measurements of Cement-based Materials (2018), ISBN: 978-2-35158-230-5, e-ISBN: 978-2-35158-231-2; *Eds. Chafika Djelal, Yannick Vanhove*

PRO 133 (online version): CO2STO2019—International Workshop CO2 Storage in Concrete (2019), e-ISBN: 978-2-35158-232-9; *Eds. Assia Djerbi, Othman Omikrine-Metalssi, Teddy Fen-Chong*

PRO 134: 3rd ACF/HNU International Conference on UHPC Materials and Structures - UHPC'2020, ISBN: 978-2-35158-233-6, e-ISBN: 978-2-35158-234-3; *Eds. Caijun Shi & Jiaping Liu*

RILEM Reports (REP)

Report 19: Considerations for Use in Managing the Aging of Nuclear Power Plant Concrete Structures (ISBN: 2-912143-07-1); *Ed. D. J. Naus*

Report 20: Engineering and Transport Properties of the Interfacial Transition Zone in Cementitious Composites (ISBN: 2-912143-08-X); *Eds. M. G. Alexander, G. Arliguie, G. Ballivy, A. Bentur and J. Marchand*

Report 21: Durability of Building Sealants (ISBN: 2-912143-12-8); *Ed. A. T. Wolf*

Report 22: Sustainable Raw Materials—Construction and Demolition Waste (ISBN: 2-912143-17-9); *Eds. C. F. Hendriks and H. S. Pietersen*

Report 23: Self-Compacting Concrete state-of-the-art report (ISBN: 2-912143-23-3); *Eds. Å. Skarendahl and Ö. Petersson*

Report 24: Workability and Rheology of Fresh Concrete: Compendium of Tests (ISBN: 2-912143-32-2); *Eds. P. J. M. Bartos, M. Sonebi and A. K. Tamimi*

Report 25: Early Age Cracking in Cementitious Systems (ISBN: 2-912143-33-0); *Ed. A. Bentur*

Report 26: Towards Sustainable Roofing (Joint Committee CIB/RILEM) (CD 07) (e-ISBN 978-2-912143-65-5); *Eds. Thomas W. Hutchinson and Keith Roberts*

Report 27: Condition Assessment of Roofs (Joint Committee CIB/RILEM) (CD 08) (e-ISBN 978-2-912143-66-2); *Ed. CIB W 83/RILEM TC166-RMS*

Report 28: Final report of RILEM TC 167-COM 'Characterisation of Old Mortars with Respect to Their Repair (ISBN: 978-2-912143-56-3); *Eds. C. Groot, G. Ashall and J. Hughes*

Report 29: Pavement Performance Prediction and Evaluation (PPPE): Interlaboratory Tests (e-ISBN: 2-912143-68-3); *Eds. M. Partl and H. Piber*

Report 30: Final Report of RILEM TC 198-URM 'Use of Recycled Materials' (ISBN: 2-912143-82-9; e-ISBN: 2-912143-69-1); *Eds. Ch. F. Hendriks, G. M. T. Janssen and E. Vázquez*

Report 31: Final Report of RILEM TC 185-ATC 'Advanced testing of cement-based materials during setting and hardening' (ISBN: 2-912143-81-0; e-ISBN: 2-912143-70-5); *Eds. H. W. Reinhardt and C. U. Grosse*

Report 32: Probabilistic Assessment of Existing Structures. A JCSS publication (ISBN 2-912143-24-1); *Ed. D. Diamantidis*

Report 33: State-of-the-Art Report of RILEM Technical Committee TC 184-IFE 'Industrial Floors' (ISBN 2-35158-006-0); *Ed. P. Seidler*

Report 34: Report of RILEM Technical Committee TC 147-FMB 'Fracture mechanics applications to anchorage and bond' Tension of Reinforced Concrete Prisms—Round Robin Analysis and Tests on Bond (e-ISBN 2-912143-91-8); *Eds. L. Elfgren and K. Noghabai*

Report 35: Final Report of RILEM Technical Committee TC 188-CSC 'Casting of Self Compacting Concrete' (ISBN 2-35158-001-X; e-ISBN: 2-912143-98-5); *Eds. Å. Skarendahl and P. Billberg*

Report 36: State-of-the-Art Report of RILEM Technical Committee TC 201-TRC 'Textile Reinforced Concrete' (ISBN 2-912143-99-3); *Ed. W. Brameshuber*

Report 37: State-of-the-Art Report of RILEM Technical Committee TC 192-ECM 'Environment-conscious construction materials and systems' (ISBN: 978-2-35158-053-0); *Eds. N. Kashino, D. Van Gemert and K. Imamoto*

Report 38: State-of-the-Art Report of RILEM Technical Committee TC 205-DSC 'Durability of Self-Compacting Concrete' (ISBN: 978-2-35158-048-6); *Eds. G. De Schutter and K. Audenaert*

Report 39: Final Report of RILEM Technical Committee TC 187-SOC 'Experimental determination of the stress-crack opening curve for concrete in tension' (ISBN 978-2-35158-049-3); *Ed. J. Planas*

Report 40: State-of-the-Art Report of RILEM Technical Committee TC 189-NEC 'Non-Destructive Evaluation of the Penetrability and Thickness of the Concrete Cover' (ISBN 978-2-35158-054-7); *Eds. R. Torrent and L. Fernández Luco*

Report 41: State-of-the-Art Report of RILEM Technical Committee TC 196-ICC 'Internal Curing of Concrete' (ISBN 978-2-35158-009-7); *Eds. K. Kovler and O. M. Jensen*

Report 42: 'Acoustic Emission and Related Non-destructive Evaluation Techniques for Crack Detection and Damage Evaluation in Concrete'—Final Report of RILEM Technical Committee 212-ACD (e-ISBN: 978-2-35158-100-1); *Ed. M. Ohtsu*

Report 45: Repair Mortars for Historic Masonry—State-of-the-Art Report of RILEM Technical Committee TC 203-RHM (e-ISBN: 978-2-35158-163-6); *Eds. Paul Maurenbrecher and Caspar Groot*

Report 46: Surface delamination of concrete industrial floors and other durability related aspects guide—Report of RILEM Technical Committee TC 268-SIF (e-ISBN: 978-2-35158-201-5); *Ed. Valerie Pollet*

Contents

1   **Introduction** 1
    Eddie Koenders, Kei-ichi Imamoto, and Anthony Soive

2   **Models for Chloride Ingress—An Overview** 7
    Lars-Olof Nilsson

3   **Marine Submerged** 25
    Anthony Soive, Neven Ukrainczyk, and Eddie Koenders

4   **Road Sprayed** 59
    Jan Bisschop and Lars-Olof Nilsson

5   **Conclusions** 81
    Eddie Koenders, Kei-ichi Imamoto, and Anthony Soive

Chapter 1
Introduction

Eddie Koenders, Kei-ichi Imamoto, and Anthony Soive

Abstract Models to predict the chloride ingress into the concrete cover of existing or new concrete structures are available in various levels of complexity with all having often their own peculiarities. This chapter introduces the most relevant motivations, definitions, and limitations that have led to the current RILEM state of the art report (STAR) on benchmarking chloride ingress modelling. Background information on the various chloride ingress models, definitions on the diffusion coefficient, benchmarking information, and the basic motivation for the RILEM TC 270-CIM are the main topics addressed in this chapter as an introduction to this STAR.

1.1 Chloride Ingress Models

Modelling the ingress of chlorides into a concrete cover is a phenomenon that has already been studied for many decades. However, many questions still remain unanswered that are related to the way chlorides accommodate and/or being transported through a concretes' capillary pore system. Over the years, many different models have been developed for predicting the diffusion process of chlorides process, with all having their own particular uncertainties among the simulated results. Many research groups around the world are still heavily involved in this topic where all are employing their own analytical and/or numerical simulation model to assess the ingress of chlorides with time. The prediction accuracy of these so called "Chloride Ingress Models" (CIM) is still a very important point since it directly relates to the actual condition and related service life assessment of a concrete structure. In other words, chloride

E. Koenders (✉)
Technical University of Darmstadt, Darmstadt, Germany
e-mail: koenders@wib.tu-darmstadt.de

K. Imamoto
Tokyo University of Science, Tokyo, Japan

A. Soive
Cerema, Aix-en-Provence, France

© RILEM 2022
E. Koenders et al. (eds.), *Benchmarking Chloride Ingress Models on Real-life Case Studies—Marine Submerged and Road Sprayed Concrete Structures*,
RILEM State-of-the-Art Reports 37, https://doi.org/10.1007/978-3-030-96422-1_1

ingress models can be employed to assess the moment that the chloride concentration at the rebar would exceed a defined critical threshold, which unleashes an irreversible deterioration of the concrete rebar corrosion. It is the moment in time after which the durability of a concrete structure reaches the point where maintenance and repair are very likely to evolve, leading to significant cost enhancements. Therefore, the more accurate the actual condition of a concrete structure can be assessed, the better the service life can be predicted and the related maintenance costs be quantified.

Models that can be employed for predicting the ingress of chlorides into a concrete surface with time are mostly based on Fick's first or second law of diffusion, which is a partial differential equation that can be solved analytically or numerically. In the daily practice the analytical solution is mostly used because of its simple form and its ease to employ in regular concrete applications. The numerical method solves the differential equation in a discretized way according to e.g. the finite difference method, finite volume method or finite element method. The analytical method has some limitations on geometry and boundary conditions, making it only applicable for a one dimension semi-infinite medium, with a constant apparent diffusion coefficient in inward direction, while exposed to a constant chloride surface condition. Moreover, the solution is applicable for a fully water saturated pore system at constant humidity and temperature. A numerical solution instead, has much more flexibility in terms of boundary conditions, starting conditions, space dimensions, etc. In particular, a numerical method can be applied to calculate the chloride ingress for any geometry and any variation of the surface chloride concentration with time. This means that numerical models are more suitable for non-constant surface chloride exposure situations than analytical models. In addition, a numerical model also provides the ability to take into account many other local conditions as well. Both the analytical and numerical approach consider the diffusion of chloride ions through the capillary pore system as the main mechanism for chloride ingress. Diffusion is a phenomenon where ionic species move through an aqueous medium while driven by concentration gradients. According to an official (chemistry) definition (merriam-webster.com) it is "the process whereby particles of liquids, gases, or solids interact as the result of their spontaneous movement caused by thermal agitation and in dissolved substances move from a region of higher to one of lower concentration". From this definition, it becomes clear that, (from a Modeller's point of view), for chloride ingress models at least the following three items are needed: an aqueous medium; a diffusing species; and a driving force. Mapping this situation to the situation that prevails in a concrete cover results in the following associating components, which are, pore water (aqueous medium), chloride ions (diffusing species) and the other ions with which they can react, and chloride concentration gradient (driving force). It makes implicitly clear that the pore water is the medium that facilitates the movement of chlorides where also the pore wall area should be considered, because of chemical reactions between ionic and solid species (chloride binding). This also means that without pore water (dry pores) there can be no chloride diffusion. This makes it more complicated since for those pores, which are partly saturated, or even where pore humidity comes into play, is making the chloride ingress a much more complex mechanism. It leads to a situation where also pore surface interactions, or more generally, carbonation reactions may affect chloride movement to a large extend, often expressed in terms of chloride binding. From this, the chloride ions that are occurring in a porous system of cement-based materials can be subdivided into two components in-which they may locally appear. This is the situation of where the total chlorides are subdivided in free chlorides and bound chlorides. Schematizing this situation, where the total chloride concentration C_t is the summation of the free chloride concentration c_f and bound chloride concentration C_b, this leads to $C_t = c_f + C_b$. Such decomposition into free and bound chlorides is also a way to differentiate between the ability to use an analytical or numerical model, where the first one tries to solve the total chloride amount of chloride diffusing into a concrete cover, and the latter is usually aiming at predicting the diffusion of the free chlorides only. From an experimental point of view, there is consensus that the amount of measured chlorides are the total chlorides, rather than the free chlorides. Hence, numerical models must use or estimate a chloride binding isotherm that describes the relationship between bound and free chlorides. The way how the free and bound chlorides are handled in the analytical and numerical models is mostly defined by the diffusion coefficient. In most cases the definition of the diffusion coefficient should cope with an implicit or explicit formulation of how the free and bound chlorides are considered and how the models separate between these two chloride concentrations. Therefore, in the next section a list of definitions on the various types of the diffusion coefficients, also used in the benchmark, are reported and explained in more detail.

1.2   Definitions

Models that can be employed for simulating the ingress of chlorides into a concrete cover generally contain numerous input parameters that are depending on their level of detail and complexity. Most important one is the diffusion coefficient of the concrete cover, which represents the accessible pore space for chlorides to enter the inner concrete and is specified for each model in a predestined way. As various models with different complexities are employed in this STAR, it is important to exactly understand the definition of each diffusion coefficient employed in a certain model, in order to also know the way how the free and bound chlorides are considered. The different diffusion coefficients, and their corresponding definitions are, therefore, described in more detail in Table 1.1.

   In general, the diffusion coefficient is determining the diffusivity of the dissolved free chloride ions in the pore water under fully saturated conditions. Any change of it will directly affect the final result of the chloride ingress simulation. The various diffusion coefficients and their definitions employed in this STAR are described in Table 1.1. A more extensive description of the various models, including details on the mathematical backgrounds and diffusion coefficients is provided in Chap. 2.

Table 1.1 Overview of the various diffusion coefficient definitions

Symbol	Definition	Description
D_{F1}	Diffusion coefficient in Fick's first law	Basic diffusion coefficient in Fick's first law differential equation (related to free concentration c_f and effect of pore morphology)
D_{F2}	Diffusion coefficient in Fick's second law	Basic diffusion coefficient in Fick's second law differential equation (related to total concentration C_t)
$D_{av,ex}$	Apparent diffusion at the time of exposure	Apparent diffusivity which is deduced from a total chloride profile at the time of exposure
D_{av}	Averaged time dependent apparent diffusion coefficient	$D_{av,ex}$ multiplied with a time function. The average of D_{F2} in the time interval $0 - t$
D_{RCM}	Migration coefficient	Measured with a migration test (RCM)
D_i	Diffusion coefficient	Diffusion coefficient (D_{F1}) for ionic transport of species i

1.3 Benchmarking Chloride Ingress

In the scientific community, but also in the daily practice, there are still discussions going on about the impact and uncertainties the various model parameters may have on the final simulated chloride ingress result. Reason for this is that in particular the diffusion coefficient and the input parameter called "ageing factor" may strongly affect the rate of the chloride ingress, and with this, may have a large impact on the prediction accuracy of the service life assessment of concrete structures. In particular for the ageing factor a sound physical background is lacking. It can, therefore, be considered as a kind of parameter that can be used to calibrate analytical chloride ingress models, to any kind of chloride affected concrete structure. This means that for existing structures, this factor could be determined quite accurately, as it could possibly be calibrated on chloride profiles received from samples taken from a chloride induced concrete structure. On the contrary, for new structures, this possibility does not exist, which is the reason why this factor is still under strong discussions since a clear calibration strategy is lacking, leading to unreliable long-term predictions. In some cases, input values may also be "assessed", or based on expert judgements. However, this does not really reduce the extreme uncertainties in the chloride ingress model predictions, and also in the probabilistic service-life assessment simulations. It is one of the reasons why a technical committee was proposed to RILEM with the aim to compare the prediction performance of analytical and numerical models for chloride ingress, leading to the proposal of RILEM TC 270-CIM. Main task of the committee was to benchmark most widely used analytical and numerical models for chloride ingress, comparing the results, and providing recommendations for possible future calibration strategies. The models can be either engineering- or scientifically-based and may cover the design and/or condition assessment stage of concrete structures exposed to chlorides originating mostly from different exposure conditions, such as marine, road or near shore conditions. Two typical benchmark cases with which the salt exposure condition at the surface of concrete structures can be characterized are a *continuous* surface concentration indicated as "Marine submerged" and a *non-continuous* surface concentration, indicated as "Road sprayed". In this STAR, therefore, the first benchmark case is indicated as *Marine submerged* where both analytical and numerical approaches are considered, whereas the second case is indicated as *Road sprayed* where mainly numerical models are considered because of the non-continuous exposure of the surface chlorides concentration with time, caused mostly by deicing salts, sprayed on a concrete structures' surface by passing traffic during Winter. Benchmarking the two mentioned cases provides the ability to receive systematic information and dependent influences of the performance of various analytical or numerical models to assess the chloride ingress into the cover of a concrete structure, where the experimental benchmark data has been received from destructive field tests. The proposed benchmark cases may serve as a reference for present and future generations of chloride ingress models, and can provide academics, engineers and/or asset owners an instrument to evaluate and/or calibrate the performance of such models used for service-life assessment of concrete structures.

In this STAR, a total of 14 different types of models from various research groups all around the world have participated in benchmarking the Marine submerged and Road sprayed cases. These models ranged from simple analytical models to very complex numerical models. Main differences were mostly on the complexity of the differential equation describing the chloride ingress mechanism by either diffusion, convection, or ionic multi-species transport, or by a combination of these three possible ways of chloride transport. Moreover, the solution strategy, i.e. analytical or numerical, differences in boundary conditions, ageing factor, diffusion coefficient, surface concentration and chloride binding are some of the features that were treated differently in the various models of the participating research groups. Due to this wide spread of possible modelling approaches and type of models employed around the world, a deeper understanding on the effect of various model parameters on chloride ingress predictions was needed, and was materialized in the technical committee TC-270 CIM.

1.4 Technical Committee TC 270-CIM

The RILEM technical committee TC 270-CIM was active between 2016 and 2021, with the main objective to establish a STAR and recommendations by benchmarking various analytical and numerical chloride ingress models for two representative cases, i.e. Marine submerged and Road sprayed. The scope was limited to only chloride ingress modelling, while not considering related subjects like rebar corrosion, carbonation, cracks, etc., leading to a situation where only the performance of the chloride ingress models themselves was analysed and compared. The TC structured its STAR as follows; following this introduction, in Chap. 2 an overview is provided of the various chloride ingress models that were employed in the benchmarking, while classifying them as empirical and physical based models. In Chap. 3, the Marine submerged benchmark results are reported where data was taken from an experiment that took place in the North seawater, on the west coast of Sweden, the so-called "Träslövsläge field experiment". This experiment comprehended various long term in situ chloride ingress measurements on concrete blocks permanently submerged in chloride containing water, representing a constant surface chloride concentration. Chapter 4, reports the Road sprayed benchmarking results where the data was taken from a road-side field experiments in Bonaduz, Switzerland, where concrete test panels alongside a road were exposed to chlorides from de-icing salts, sprayed by passing traffic during winters. This test case was selected to represent a non-constant surface chloride concentration, which required from the chloride ingress models a different handling of the surface concentration with time. Finally, Chap. 5 reports the recommendations and conclusions, where also the next possible steps for chloride ingress model calibrations are addressed.

Chapter 2
Models for Chloride Ingress—An Overview

Lars-Olof Nilsson

Abstract The available chloride ingress models are described in a structured way, separating empirical models from physical models. Empirical models are based on fitting measured data to an equation and deriving some parameters that are then used for predictions. Physical models use flux equations and descriptions of equilibrium conditions between the species in the pore solution and the matrix. Almost all models can be grouped into these two categories. The following description is adopted from (Chlortest, 2006) and (Nilsson, 2011). The presentation starts with a few general remarks on concrete in chloride ingress models and on Fick's laws. Then various chloride ingress models are described in principle, starting with empirical models, followed by different physical models using Fick's first law or the Nernst-Planck-equation as flux descriptions.

2.1 Model Concrete Versus Real Concrete

Almost all chloride ingress models, also very sophisticated ones, describe concrete and the conditions in it in very simplified ways, especially when it comes to the effects of self-desiccation, the inhomogeneity of concrete, the continuous binder reactions and various surface effects.

2.1.1 Self-desiccation and Air Voids

All concretes that are relevant in environments with chloride exposure, i.e. low-w/b-concretes, have significant self-desiccation. That means that the pore system is partly emptied at early ages. This empty part of the pore system will eventually be filled with chloride containing water which adds to the chloride that penetrated due to

L.-O. Nilsson (✉)
Moistenginst AB, Trelleborg, Sweden
e-mail: lars-olof.nilsson@byggtek.lth.se

© RILEM 2022

E. Koenders et al. (eds.), *Benchmarking Chloride Ingress Models on Real-life Case Studies—Marine Submerged and Road Sprayed Concrete Structures*,
RILEM State-of-the-Art Reports 37, https://doi.org/10.1007/978-3-030-96422-1_2

diffusion. This effect is not small. An estimate is the once called "contraction pores" of some 0.06 α litres per kg of cement. With a degree of reaction of $\alpha = 0.6$, this means some 0.04 L per kg of cement. With a chloride content of 20 g/l in sea water, that means an additional chloride content of 0.8 g/kg cement or 0.08% by weight. Surface chloride contents in concrete exposed to sea water is around 3–6% by weight of cement. Neglecting the effect of self-desiccation may describe only some 2% of the chloride in a chloride profile as being transported by diffusion when the true ingress mechanism is convection with the ingress of sea water.

In the Northern countries concrete exposed to chloride is given an air void system in order to protect the concrete against frost damage. These air void systems may be some 4–6% by volume of concrete and are supposed to be air filled during the entire service-life of the concrete. Some of the smaller air voids, however, may be filled with water with time, adding to the volume of pore water that can contain chlorides.

2.1.2 Concrete Inhomogeneity

In "model concrete" the effect of the aggregate is usually not regarded at all. Model concrete is almost always considered a fully homogenous material with a certain binder content (kg/m^3). This may be appropriate since the properties that are determined in various tests are measured on the same material that has the same binder/aggregate content and presence of microcracks.

Close to a cast concrete surface, however, the binder content is much higher than in the bulk concrete due to the wall effect. Large aggregate grains cannot be present very close to the formwork. Consequently, the binder content is higher the closer to the surface.

This effect is not included in any chloride ingress model. In chloride profile measurements, however, the chloride content will be higher closer to the surface because of this wall effect. When predicted chloride profiles are compared to early exposure data this comparison is significantly distorted by the effect of this "inhomogeneity" of the binder content.

2.1.3 Continuous Binder Reactions

Concrete is usually fairly young when first exposed to sea water, may be only one or two weeks old. That means that the concrete will continue to mature in parallel with the ingress of chloride. Most of the self-desiccation happens within the first month emptying part of the pore system. That means that the binder reactions will start to slow down at an early age in contrary to the surface regions where the pore system is likely to be more or less saturated with water. The maturity of the concrete consequently can be very different at different depths and these conditions change over time.

2.1.4 Surface Effects

The surface concrete is different from the bulk concrete for many other reasons as well. Sea water does not only contain chlorides but also a number of other species, some of which will react with binder constituents. The concrete surface maybe carbonated and ions will leach out of the concrete.

2.2 Fick's Laws and the Two Diffusion Coefficients

Most models for chloride ingress are solutions to Fick's 2nd law.

$$\frac{dC}{dt} = D_{F2}\frac{d^2C}{dx^2} \tag{2.2.1}$$

where C is the total amount of chloride (kg/m^3), t is time (s).

D_{F2} is the diffusion coefficient, or more correctly, the diffusivity in Fick's 2nd law (m^2/s),

x is the Depth Position (m).

This equation looks like a mass balance equation, but a mass balance equation should contain a term describing the transport of chloride and the total amount of chloride C is not a proper transport potential. Diffusion of chlorides should use the concentration of free chloride c (kg/m^3 liquid) as the transport potential.

The flux of chlorides due to diffusion is then described by Fick's 1st law (of diffusion)

$$J = -D_{F1}\frac{dc}{dx} \tag{2.2.2}$$

where J is the flux of chlorides (kg/m^{2}s),

D_{F1} is the diffusion coefficient (m^2/s).

The mass balance of chlorides in a volume of concrete with a thickness of dx is obvious from Fig. 2.1.

Assuming constant properties one gets

$$\frac{dC}{dt} = -\frac{dJ}{dx} = -\frac{d}{dx}\left(-D_{F1}\frac{dc}{dx}\right) \tag{2.2.3}$$

This can be rearranged to free and bound chlorides by

Fig. 2.1 The mass-balance for chlorides in an infinitesimal volume of concrete with a thickness of dx

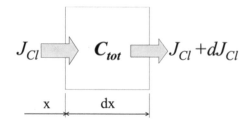

$$\frac{dC}{dt} = \frac{d}{dx}\left(\frac{D_{F1}}{\frac{dC}{dc}}\frac{dC}{dx}\right) \tag{2.2.4}$$

Or, if D_{F1} and dC/dc are constants in space

$$\frac{dC}{dt} = \frac{D_{F1}}{\frac{dC}{dc}}\frac{d^2C}{dx^2} \tag{2.2.5}$$

Comparing this mass-balance equation with Fick's 2nd law, Eq. (2.2.1), the relationship between the two diffusion coefficients is identified

$$D_{F2} = \frac{D_{F1}}{\frac{dC}{dc}} \tag{2.2.6}$$

where

dC/dc is the chloride binding capacity with the dimension.
(kg Cl/ m^3 concrete / kg Cl /m^3 liquid).

The two diffusion coefficients have the same dimension (m^2/s) but they differ depending on the porosity of the concrete and the chloride binding properties of the concrete binder.

2.3 ERFC-Solution with Constant D$_{F2}$

The solution to Fick's 2nd law (Eq. (2.3.1)), with a constant D_{F2}, if the initial chloride content is assumed to be zero, reads

$$C(x,t) = C_s \cdot \mathrm{erfc}\left(\frac{x}{2\sqrt{D_{F2} \cdot t}}\right) \tag{2.3.1}$$

where

$C(x,t)$ is the total chloride content at depth x (m) at exposure time t (s),

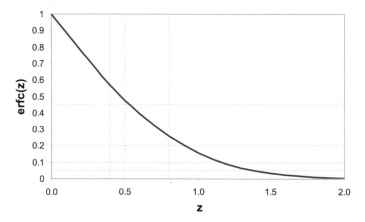

Fig. 2.2 The error-function complement erfc(z) where $z = x/2\sqrt{Dt}$

C_S is the constant surface chloride content at depth $x = 0$.

The erfc equation is shown in Fig. 2.2.

2.4   ERFC-Solution with Time-Dependent D_{F2}

The diffusivity D_{F2} will of course not be a constant but decreasing with time. The solution to Fick's 2nd law will still be the same, with the constant D_{F2} replaced by the average value of D_{F2} in the time interval $(0 - t)$.

$$C(x, t) = C_s \cdot \text{erfc}\left(\frac{x}{2\sqrt{D_{F2} \cdot t}}\right) \tag{2.4.1}$$

The average of D_{F2} in the time interval $0 - t$ is called the "average" diffusivity D_{av}.

$$C(x, t) = C_s \cdot \text{erfc}\left(\frac{x}{2\sqrt{D_{av} \cdot t}}\right) \tag{2.4.2}$$

This equation also contains the chloride surface content C_s. This is the mostly used equation to analyze data from specimens or structures that have been exposed to chloride during an exposure time t. Curve-fitting data-points to this equation gives what diffusivity D_{av} and chloride surface content C_{sa} that are "achieved" or "apparent" after this time.

Note that the diffusivity in Fick's 2nd law D_{F2} and the average diffusivity D_{av} are different; the latter is the average of the former.

A common way to express the time-dependency of the average diffusivity is this equation

$$D_{av} = D_{av,ex} \cdot \left(\frac{t_{ex}}{t}\right)^{\alpha} \qquad (2.4.3)$$

where t is the age of the concrete, which most often is very close to the time of exposure, and t_{ex} is the age of the concrete at first exposure. $D_{av,ex}$ and α are parameters expressing the apparent diffusivity at the time of exposure and the age exponent, respectively. We have no other way to determine these two parameters than curve-fitting chloride profiles from several exposure times to Eq. (2.4.2).

The so called "DuraCrete Model" is using an expression similar to Eq. (2.4.3) with a D_{RCM} from a migration test as input and with a number of "correction factors" for considering the effects of test method, environment and curing.

$$D_{av} = k_t \cdot D_{RCM} \cdot k_e \cdot k_c \cdot \left(\frac{t_{ex}}{t}\right)^{\alpha}$$

Later, the DuraCrete model was modified to exclude a convection zone with a certain thickness Δx, since most chloride profiles have a peak close to the surface. The chloride surface content is replaced with a chloride content C_s, Δx at the depth of the peak

$$C(x, t) = C_{s,\Delta x} \cdot \mathrm{erfc}\left(\frac{x - \Delta x}{2\sqrt{D_{av} \cdot t}}\right) \qquad (2.4.4)$$

The effect of neglecting the concrete surface region with a thickness of Δx has not been clearly quantified. The effect should be significant at shorter exposure times.

2.5 The False Erfc-Solution with Time-Dependent D_{F2} and C_s

A simple chloride ingress model with time-dependent diffusivity Da and time-dependent $C_s(t)$ is the "false erfc-solution". This is done by using Eq. (2.4.2) but expressing the chloride surface content as time dependent, $C_s(t)$. Such an equation is of course possible to use for curve-fitting data-points and get results that seems to be able to describe the data.

Note, however, that such an equation is not a mathematically correct solution to Fick's 2nd law. It must not be used at all, and certainly not for prediction beyond test dates.

2.6 The Mejlbro-Poulsen Model—Time-Dependent D_{F2} and C_s

The correct solution to Fick's 2nd law with time-dependent D_{F2} and C_s was derived by Mejlbro in 1996 (Mejlbro, 1996).

$$C(x, t) = C_s(t) \cdot \Psi_p\left(\frac{x}{2\sqrt{D_{av} \cdot t}}\right) \qquad (2.6.1)$$

where Ψ_p is the Mejlbro-Poulsen function, see below.

In order to be able to find an analytical solution at all, the time-dependent chloride surface content $C_s(t)$ had to be described in a fairly complicated way.

$$C_s(t_{ex}, t) = C_i + S \cdot \left((t - t_{ex}) \cdot D_{av,ex} \cdot \left(\frac{t_{ex}}{t}\right)^\alpha\right)^p \qquad (2.6.2)$$

where C_i is the initial chloride content. S and p are the two new parameters. The other parameters are the same as in Eqs. (2.4.2) and (2.4.3) (Fig. 2.3).

With this way of describing the chloride surface content $C_s(t)$, Mejlbro-Poulsen (Mejlbro, 1996) found the solutions Ψ_p to Fick's 2nd law with time-dependent $D_{av}(t)$ and $C_s(t)$. Solutions for a number of values of the parameter p are shown in Fig. 2.4.

Analytical solutions to Fick's 2nd law have a large number of parameters. The relationship between all these parameters are frequently causing a lot of confusion. Evaluation of parameters from field data using different solutions is sometimes

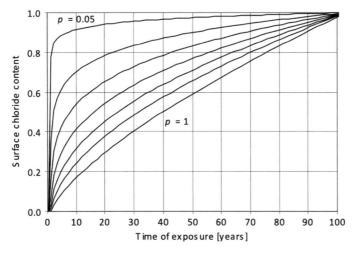

Fig. 2.3 Alternative ways to describe the time-dependency of the chloride surface content $C_{sa}(t)$ in the Mejlbro-Poulsen model (Mejlbro, 1996)

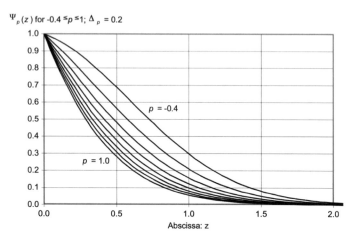

Fig. 2.4 The Mejlbro-Poulsen solutions to Fick's 2nd law for different values of the parameter p. A value of $p = 0$ gives the erfc-solution (Mejlbro, 1996)

erroneous because of this confusion. The confusion was completely clarified by Frederiksen et al. (2008).

2.7 Numerical Solutions to Fick's 2nd Law

Fick's 2nd law can of course be treated as a mass-balance equation for chlorides, with the total chloride content C as the chloride transport potential and the diffusivity D_{F2} as the chloride transport coefficient. The equation can be easily solved numerically even if the diffusivity and the chloride surface contents are time dependent. A number of chloride ingress models are using this principle.

The diffusivity $D_{F2}(t)$ is, however, difficult to quantify. Since it is not the same as the apparent diffusivity D_a, cf. Equations (2.4.1) and (2.4.2), there is an obvious risk that the diffusivity D_{av} is erroneously used instead.

2.8 The Square Root Model

Already in 1985, Poulsen et al. (1985) described depth ingress, x_{cr}, of a certain "reference chloride content" as a function of a simple square-root of time. "Reference chloride content" is usually the critical chloride content for corrosion initiation C_{cr}.

$$x_{cr} = K \cdot \sqrt{t} \qquad (2.8.1)$$

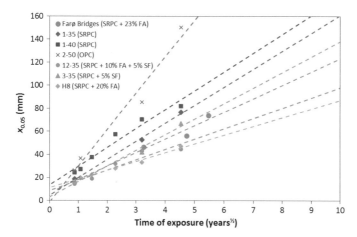

Fig. 2.5 Field data points of penetration depths of a chloride content corresponding to 0.05% by weight of concrete from seven concretes exposed to sea water for up to 30 years compared to the square root model (lines) (Poulsen & Sørensen, 2014)

where K is a parameter to calculate. This equation actually follows from Eq. (2.3.1), by using the inverse of erfc, $erfc^{-1}$.

Recently, this model has been further developed and compared to a lot of field data (Poulsen & Sørensen, 2014). It has been found that the model needs an offset b_C at $t = 0$ to account for a number of early processes in concrete before and after exposure.

$$x_{cr} = b_C + a_C \cdot \sqrt{t} \qquad (2.8.2)$$

a_c and b_c are fit parameters to be calculated from the data. With such a model there is no need for any time-dependency of the parameters. Examples of the square-root of time model compared to field data is shown in Fig. 2.5.

2.9   Physical Models—General Aspects

"Physical models" for chloride ingress are models where the physical (and chemical) processes that influence chloride ingress are described as correct as possible. These models solve the mass balance equations for chlorides and other ions by using ion transport equations and equations for the interaction between the ions and the matrix. In doing so, some models are limited to chlorides while others include some or several other ions as well, in a multi-species approach. Very advanced physical models are describing this transport and interaction under non-saturated conditions, e.g. are including non-saturated ion transport and convection of ions.

Consequently, all physical models for chloride ingress must describe these two types of processes:

- transport of chlorides, and other ions,
- interaction between the ions and between the ions and the matrix.

Examples of models for these processes are described below. A large number of physical models are developed where in these models the individual processes are combined in different ways.

2.10   Physical Models—Ion Transport Models

Some physical models for chloride ingress use Fick's 1st law, cf. Eq. (2.2.2), to describe diffusion of chlorides. Sometimes transport of some other ions are described in parallel, also with Fick's 1st law. This is of course a simplification since the interaction with the other ions in the pore solution is neglected.

A multi-species approach is frequently used to describe the transport of several ions with the Nernst-Planck equation. The flux of ion i is described as follows (see Eq. 2.10.1):

$$J_i = -D_{F1,i}\left(\frac{\partial c_i}{\partial x} + c_i\frac{\partial lna_i}{\partial x} + \frac{z_i F}{RT}c_i\frac{\partial \Phi}{\partial x}\right) \tag{2.10.1}$$

where

J_i is the flux of ion i (kg/m^{2}s) or (mol/m^{2}s),
$D_{F1,i}$ is the diffusion coefficient of ion i (not the diffusivity) (m^2/s),
c is the concentration (kg/m^3 liquid) or (mol/m^3 liquid),
a is the activity (−),
Φ is the electrical potential (V),
F is the Faraday constant.

Using the Nernst-Planck equations for describing the flux of ions in the model may require a parallel description of the electrical field as well.

2.11   Physical Models—Interaction Between Ions and Matrix

The interaction between one or more ions and the matrix can be described in various ways in physical models for chloride ingress. The simplest way is to use a chloride "binding isotherm". Such an isotherm is a relationship between the bound or total amount of chloride in a unit volume of concrete and the concentration of chlorides in the pore solution. An example is given in Fig. 2.6.

Fig. 2.6 An example, in principle, of a chloride binding isotherm

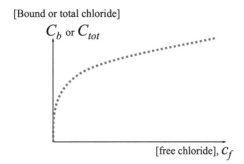

The term binding <u>isotherm</u> is used because the interaction is temperature dependent. The chloride binding isotherm also depends on the pH. Consequently, the temperature variations and leaching of hydroxides should be included in a physical model where the interaction between chlorides and the matrix is described with a chloride binding isotherm.

Chloride "binding" is also affected by a number of other ions in the pore solution. This interaction cannot be fully described by a "binding isotherm". Instead, the equilibrium conditions for each species must be quantified in a multi-species model. This is done, for example, in the software Stadium® (Samson & Marchand, 2007) with a special "chemical module" that quantifies the interaction between the individual species and the various solid phases in the binder. A similar approach was used by Johannesson et al. (2007).

Some chloride ingress models are using a software tool, PHREEQC, for calculating these equilibrium conditions. This tool is actually freely available from www.usgs.gov. GEMS could be used for the same purpose (Loser et al., 2010).

2.12 Physical Models—Non-saturated Conditions

Non-saturated conditions require physical models for chloride ingress where the non-saturated ion transport by diffusion and convection can be correctly described. The diffusion coefficient D_i is described as a function of the pore humidity RH or the degree of saturation S. For describing convection of ions, moisture transport has to be divided into two parts: one that can carry ions and one that cannot. This division is still a topic for research.

How the interaction between the ions in the pore solution and the matrix should be described under non-saturated conditions is also not clarified. In some models the same chloride binding isotherm is used as for saturated conditions, cf. Fig. 2.6, with the chloride bound to the matrix as a function of the concentration of free chloride in the pore water that does not completely fill the pores. Such an approach is not yet validated by direct measurements.

2.13 Boundary Conditions

The boundary condition in empirical models or models based on Fick's 2nd law in general, is the surface chloride content C_s. For concrete submerged in sea water C_s could be calculated from the chloride concentration c_f in the sea water and the chloride binding isotherm for the concrete. Such a calculation, however, is not simple since the chloride binding is significantly temperature dependent. The C_s depends more on sea water temperature than on the salinity (Lindvall, 2003).

For other environments C_s must be quantified from field data, considering a large number of environmental parameters like distance, horizontally and vertically to a chloride source like sea water table and road surface with deicing salts. The orientation and roughness of the concrete surface matters as well together with a number of other parameters.

Boundary conditions for physical models are even more complicated, except for concrete submerged in sea water. In the tidal and splash zones and close to roads with de-icing salts at least the concentration of chloride $c_f(t)$, temperature $T(t)$ and humidity $RH(t)$ at the very surface must be described. This is really challenging. Some attempts were made by Lindvall and Nilsson (1999).

2.14 Models Used in the Work of RILEM TC 270-CIM

A large number of models are used in the test cases in the committee work. These models are summarized here with a short description of their main contents. The modeller is mentioned in parenthesis in the headline.

2.14.1 "Double-Multi" Model (Liu)

The model proposes a numerical solution in 2-D. It solves a multi-phase of mass conservation equation for multi-species, the details of which could be found in Liu et al., (2012, 2015). The model considers several phenomena:

– Heterogeneous nature of concrete (three phases: mortar, aggregate, Interfacial Transition Zone).
– Parameterization of aggregate distribution based on Fuller gradation.
– Curve-fitting to given profiles gives parameters $D_{av,k}$ ($k = 1, 2$) and C_1 (at $x = 0$).
– Temperature-dependent $D_{av}(T)$ considering activation energy at different w/b.
– Chloride binding based on linear and non-liner isotherms (Linear isotherm for simplicity based on comparison between three types of isotherms).
– Electro-coupling effects of multiple ions due to internal charge imbalance (could be ignored during the diffusion-dominated processes).

– Fixed boundary conditions; simulated semi-infinite condition by over-long geometry.
– Input parameters: aggregate shape (regular polygon), aggregate maximum/minimum radius, aggregate volume fraction (based on composition of test specimen), apparent diffusion coefficient in mortar, ITZ thickness, ratio of $D_{av,ITZ}$ and $D_{av,mor}$, water to binder ratio, activation Energy, temperature range, experimental-determined constants for isotherms, electrostatic potential, initial and boundary concentrations, etc.

2.14.2   ERFC Models (Vu, Mai-Nhu, El Farissi)

The models used are based on the fib Model (fib, 2006), which is an erfc-solution of the Fick's 2nd law, and considers the following possibilities:

– Time-dependent and temperature-dependent $D_{av}(t)$.
– Input to D_{av} from immersion test (A) (NT BUILD 443) or migration test (B) (NT BUILD 492).
– Correction factors for test method, temperature and age (with an age exponent).
– Water convection by considering a depth, Δx, above which the diffusion process is the main phenomenon.

2.14.3   SDReaM-Crete (Mai-Nhu)

The home-made model from the Industrial Technical Center (Cerib) solves the mass balance equation for chlorides with:

– Chloride diffusion from Fick's 1st law (D_{F1}).
– Chloride convection from liquid water flow.
– Time-dependent physical chloride binding to the gel.
– Time-dependent chemical chloride binding to Friedel's salt.
– Liberation of bound chloride by carbonation.

2.14.4   ERFC (Monteiro)

The model is based on an erfc-solution of Fick's 2nd law and considers a time-dependent $D_{av}(t)$ with an age exponent and a constant C_s from field data. The diffusion coefficient is determined by curve-fitting data from total chloride profiles.

2.14.5   HETEK-Model (Monteiro)

The model is based on the Mejlbro-Poulsen-solution (Mejlbro, 2016) and considers a time-dependent $D_{av}(t)$ with an age exponent and a constant C_s from field data. The diffusion coefficient is determined by curve-fitting data from total chloride profiles.

2.14.6   Mejlbro-Poulsen Model (Frederiksen)

The model uses the Mejlbro-Poulsen-solution (Mejlbro, 2016) in an Excel sheet with the following characteristics:

– Time-dependent $D_{av}(t)$ with an age exponent.
– Time-dependent surface chloride content $C_s(t)$.
– 17 coefficients quantified by regressions analysis to 20 years field data for some 40 concretes made with CEM I, fly ash and silica fume, i.e. only valid for these binders and them combined.
– The input data is solely the concrete binder composition and the design lifetime, i.e. no further input from testing of the concrete type in question.

2.14.7   Cerema Model (Soive)

The model is a physical/chemical-based model that solves the mass and energy balance equations (Soive et al., 2018; Tran et al., 2018) with:

– Ionic species diffusion from Fick's 1st law, with the same $D_{F1,i}$ diffusion coefficient for all ions (the effect of the electrical field is neglected).
– Porosity reduction from mineral dissolution/precipitation.
– Ion interactions with matrix by Toughreact-calculations.
– Moisture transport not carrying ions described by water vapour content gradients.
– Time-dependent boundary conditions: temperature, wetness, RH, ionic concentration.

2.14.8   Freund/Langmuir (Ukrainczyk, Patel, Bahman)

The physical/chemical model (Ukrainczyk and Koenders 2016) solves the mass balance equations for Cl species with:

– Ion diffusion from Fick's 1st law, with fixed D_{F1} diffusion coefficient (neglecting the time variability of pore volume and morphology).
– Porosity reduction from mineral dissolution/precipitation not considered.

- Chloride interactions with matrix by empirical (Langmuir/Freundlich) binding isotherm.
- One modeler used Langmuir binding isotherm and method of lines numerical implementation (*pdepe* in Matlab).
- Another modeler used Freundlich binding isotherm and finite volume method implemented in open source code FiPy (https://www.ctcms.nist.gov/fipy/).
- Calibration of D_{FI} and binding isotherm parameters.

2.14.9 ClinConcOrig Model (Nilsson)

The original ClincConc model (Nilsson, 2011). A numerical solution of the mass balance equation for chlorides with:

- Chloride diffusion from Fick's 1st law with D_{RCM} as parameter.
- The increase of total chloride content split into free and bound by a non-linear, pH- and T-dependent chloride binding isotherm.
- Leaching of alkalis by diffusion from Fick's 1st law.

2.14.10 ClinConcEng Model (Tang)

The ClinConc Engineering model (Tang, 2011). An erfc-solution for free chlorides, with:

- Time-dependent $D_{av}(t)$ with an age exponent.
- Input to D_{av} from a migration test (NT BUILD 492); test method factor to bridge the relationship.
- Total chloride content by adding bound chloride from a non-linear, pH- and T-dependent chloride binding isotherm.
- Leaching not considered.
- 40 parameters in total.

2.14.11 ClinConcEngSimpl Model (Nanukuttan)

A simplified ClinConc Engineering Model (Nanukuttan et al., 2013). An erfc solution for free chlorides, with:

- Time dependent $D_{av}(t)$ with an age exponent.
- Input to D_{av} from a migration test (NT BUILD 492); test method factor to bridge the relationship.
- Total chloride content by adding bound chloride from a non-linear, pH- and T-dependent chloride binding isotherm.

- whilst the model has all 40 parameters, only six of the most significant ones were utilized; these parameters can be determined independently.
- For details, see (Nanukuttan et al., 2013).

2.14.12 HETEK-Conv Model (Nilsson)

The convection model from the HETEK project (Frederiksen, 1997). A numerical solution for the mass-balance equation for chloride and moisture, with:

- Chloride diffusion from Fick's 1st law, moisture dependent diffusion coefficient.
- Moisture dependent, non-linear chloride binding isotherm.
- Chloride convection from liquid water flow, described with RH gradients with a moisture dependent permeability.
- Moisture transport not carrying ions described by water vapour content gradients.
- Moisture sorption isotherm dependent on chloride concentration.
- Time-dependent boundary conditions: temperature, wetness, RH, chloride concentration.

References

Chlortest. (2006). *Modelling of chloride ingress.* Work package 4 Resistance of Concrete to Chloride Ingress—From Laboratory Tests to In-Field Performance. EU-Project CHLORTEST G6RD-CT-2002-00855. FINAL TECHNICAL REPORT 2006.

Fib. (2006). *Model code for service life design*, Bulletin 34, International Federation for Structural Concrete (fib), Lausanne, Switzerland.

Frederiksen, J. M., Mejer, J., Nilsson, L.-O., Poulsen, E., Sandberg, P., Tang, L., & Andersen, A. (1997) *HETEK, A system for estimation of chloride ingress into concrete, Theoretical background.* The Danish Road Directorate.

Frederiksen, J. M., Mejlbro, L., & Nilsson, L.-O. (2008). *Fick's 2nd law—Complete solutions for chloride ingress into concrete—with focus on time dependent diffusivity and boundary condition.* http://www.byggnadsmaterial.lth.se Lund University, Division of Building Materials Report TVBM-3146, Lund 2008.

Johannesson, B., Yamada, K., Nilsson, L.-O., & Hosokawa, Y. (2007). Multi-species ionic diffusion in concrete with account to interaction between ions in the pore solution and the cement hydrates. *Materials and Structures, 40*, 651–665.

Lindvall, A., & Nilsson, L.-O. (1999). *Models for environmtal actions on concrete surfaces.* Duracrete Document BE95-1347/R3, March 1999.

Lindvall, A. (2003) *Chloride ingress data from field exposure at twelve different marine exposure locations and laboratory exposure.* Publication P-03:1, Dept of Building materials, Chalmers University of Technology, Göteborg.

Liu, Q. F., Easterbrook, D., Yang, J., & Li, L. Y. (2015). A three-phase, multi-component ionic transport model for simulation of chloride penetration in concrete. *Engineering Structures, 86*, 122–133.

Liu, Q. F., Li, L. Y., Easterbrook, D., & Yang, J. (2012). Multi-phase modelling of ionic transport in concrete when subjected to an externally applied electric field. *Engineering Structures, 42*, 201–213.

Loser, R., Lothenbach, B., Leemann, A., & Tuschmid, T. (2010). Chloride resistance of concrete and its binding capacity—Comparison between experimental results and thermodynamic modeling. *Cement and Concrete Composites, 32*(1), 34–42.

Mejlbro, L. (1996). The complete solution of Fick's 2nd law of diffusion with time-dependent diffusion coefficient and surface concentration. In *Durability of Concrete in marine concrete structures*. Cementa, Danderyd, Sweden.

Nanukuttan, S., Green, C., Basheer, M., Robinson, D., McCarter, J., & Starrs, G. (2013). Key performance indicators—A new approach for specifying concrete and assessing state of health of concrete structure. In D. Bjegović, H. Beushausen, & M. Serdar (Eds.), *RILEM International workshop on performance-based specification and control of concrete durability* (Vol. 89, pp. 301–308). [Proceedings 89] RILEM Publications s.a.r.l. http://www.rilem.org/gene/main. php?base=500218&id_publication=433.

Nilsson, L.-O. (2011). Modelling of chloride ingress. In L. Tang, L.-O. Nilsson, M. Basheer (Eds.). *Resistance of concrete to chloride ingress—Testing and modelling*. Taylor & Francis Group Ltd.

Poulsen, E., et al (1985) *13 concrete diseases, how they develop, proceed and are prevented (in Danish)*. SBI Betong 4, The Danish Institute of Building Research, Hørsholm, Denmark.

Poulsen, S. L., & Sørensen, H. E. (2014). *Chloride ingress in old Danish bridges*, 2nd ICDC, New Delhi, 4–6th Dec. 2014.

Samson, E., & Marchand, J. (2007). Modeling the transport of ions in unsaturated cement-based materials. *Computers & Structures, 85*(23), 1740–1756.

Soive, A., Tran, V.-Q., & Baroghel-Bouny, V. (2018). Requirements and possible simplifications for multi-ionic transport models—Case of concrete subjected to wetting-drying cycles in marine environment. *Construction and Building Materials, 164*, 799–808.

Tang, L., Nilsson, L.-O., & Basheer, P. A. M. (2011). *Resistance of concrete to chloride ingress: Testing and modelling*. Taylor&Francis, Spon Press.

Tran, V. Q., Soive, A., Baroghel-Bouny, V. (2018). Modelisation of chloride reactive transport in concrete including thermodynamic equilibrium, kinetic control and surface complexation. *Cement and Concrete Research, 110*, 70–85.

Ukrainczyk, N., & Koenders, E. (2016). Numerical model for chloride ingress in cement based materials: Method of lines implementation for solving coupled multi-species diffusion with binding. *Computations and Materials in Civil Engineering, 1*(3), 109–119.

Chapter 3
Marine Submerged

Anthony Soive, Neven Ukrainczyk, and Eddie Koenders

Abstract The first case employed for benchmarking the chloride ingress models is a marine test site in Sweden, where concrete panels are long-term (partly) submerged in sea water as part of an experimental research program for gathering chloride ingress data on concrete durability. In this test case, the concrete is continuously in a submerged condition which mimics a constant surface chloride concentration and is therefore indicated as *Marine Submerged*. This chapter reports the benchmarking performance of both analytical and numerical models where the calibration and prediction performance of the various participating models is compared and evaluated for 0.6, 1, 2, 5, 20, 50 and 100 years of exposure.

3.1 Introduction

A very well investigated concrete, in a well-known and defined marine environment, has been chosen as first test case for benchmarking chloride ingress models. One single concrete has been considered for the benchmarking in this RILEM-TC. In the present test case, the concrete is constantly submerged in seawater in field conditions, mimicking constant surface chloride boundary conditions. In such exposure conditions, several typical chloride concentration depth profiles could be observed, driven by time- and space-dependent reactive-transport phenomena. Firstly, the chloride ingress decreased with time in the first few months as well as in the following years of exposure. This phenomenon can be related to the decrease of the chloride diffusion coefficient, induced by changes in the pore structure resulting from e.g. ongoing hydration. Secondly, the binder composition can have an important influence on the chloride binding capacity of concretes. Thirdly, a "convection effect" is often observed even in saturated conditions, especially in field conditions. This effect

A. Soive (✉)
Cerema, Aix-en-Provence, France
e-mail: anthony.soive@cerema.fr

N. Ukrainczyk · E. Koenders
Institute of Construction and Building Materials, TU Darmstadt, Germany

© RILEM 2022

E. Koenders et al. (eds.), *Benchmarking Chloride Ingress Models on Real-life Case Studies—Marine Submerged and Road Sprayed Concrete Structures*,
RILEM State-of-the-Art Reports 37, https://doi.org/10.1007/978-3-030-96422-1_3

can be described as a chloride content increase in the first millimeter to a couple of centimeters depth of the material until a maximum is reached followed by a decrease deeper in the material. Fourthly, the surface concentration increases with time, $C_s(t)$. This concentration is obtained as the intersection between the ordinate axis and an extrapolation of the chloride profile (usually ignoring the chlorides profile in the convection zone).

Due to the difficulty to consider all physical–chemical phenomena during ingress of aggressive ionic species, a significant scatter between model predictions can be observed for a same case study (Tang et al., 2011). This scatter could be related to a lack of consideration of concrete execution processes resulting in heterogeneity of materials initial properties, as well as environmental chloride ingress relevant (boundary) conditions. A well-known "wall effect" is often observed in field conditions, and is caused when casting concrete against a formwork. This leads to a more porous material in the first millimeter's depth due to a decrease of aggregates quantity, which could even be worsened by poor compaction. A reduced curing period can lead to a gradient of physical–chemical properties due to competing behavior between hydration and drying at early age. In boundary conditions, temperature and salinity can vary with time along the period of exposure and can play a role in the chloride ingress rate. Other ionic species can interact with chlorides and the hydrated mineral phases contained in the material. All these effects are rarely, partially or differently considered in various models, leading to significant deviations among the model predictions.

Therefore, major goal of this benchmark is to allow for a critical evaluation of the performance and applicability of chloride ingress models on their practical usability in a particular field study condition. It must also permit to identify possible gaps in accuracy, coverage and/or knowledge cross-links. The following sections detail the field experiment (Sect. 3.2) and the benchmarking results (Sects. 3.3 and 3.4) for a fully saturated concrete under marine conditions.

3.2   Description of the Test Case

The experiment took place in North seawater in the west coast of Sweden where the "Träslövsläge field experiment" started in 1992 as part of the Nordic project BMB ("The Durability of Marine Concrete Structures"). Several concrete slabs were produced and exposed to four marine environments (submerged, tidal, splash and atmospheric zones). In the present study, only the submerged zone was subjected to a benchmarking analysis. During a several-year periods, cores were taken to accurately determine the chloride profiles (Tang, 1997). The slabs were submerged for 0.6, 1, 2, 5 and 20 years.

3.2.1 Concrete Properties

The concrete exposed to seawater has a CEM I cement, which also contained silica fume. The mix composition is shown in Table 3.1. The chemical composition of the cement and silica fume are not known. The specimens (slabs) were exposed at an age of 14 days and were kept submerged in seawater for the entire time.

Various chloride migration and apparent diffusion coefficients and diffusivities have been determined for this concrete at different ages (see Table 3.2). D_{CTH} is a migration coefficient determined by the "CTH method" (Tang, 1997), now NTBuild 492 (Nordtest, 1999), with 30 V over 50 mm for 24 h. D_{AEC} is determined in a five-week immersion test, similar to NTBuild 443 (Nordtest, 1995), in a 165 g NaCl/litre solution.

The binder content was measured as a function of depth at the different times of exposure (see Fig. 3.1). These profiles clearly underline the 'wall effect' at the concrete surface, where the binder content is higher in the first millimeters of the concrete surface. The consequences of such an observation can be multiple. The higher binder content is expected to increase the binding capacity of concrete (surface) due to higher alumino-ferite-monosulfate (AFm) phase content which chemically binds chlorides (into Friedel's and Kuzel's salts) and higher C–S–H content (which physically adsorbs chlorides). On the contrary, the porosity of this concrete surface layer is higher, thus the purely physical diffusion coefficient (D_{F1}) is higher which may lead to an increased chloride ingress rate (i.e. D_{F2}) and D_{av}). Consequences of such opposing effects are then difficult to deduce in terms of chloride binding capacity and total chloride content.

The concrete moisture content was measured after 2 year-exposure (see Fig. 3.2). It shows that the liquid saturation is higher throughout the concrete surface. However, this information has not been brought to the attention of the modelers.

Table 3.1 Composition of the selected test concrete

w/b	Cement (kg/m^3)	Silica fume (kg/m^3)	Water (kg/m^3)	Cement paste Vol-%	Binder content % by mass	Calculated density (kg/m^3)	Air Vol-%
0.40	399.0	21.0	168.0	30.4	19	2210	5.9

Cement: SRPC, CEM I. CaO in binder: 61.7% by weight. Silica fume: slurry

Table 3.2 Diffusion coefficients/diffusivity for test concrete from accelerated tests

D_{CTH} (10^{-12} m^2/s) [2]				D_{AEC} (10^{-12} m^2/s) [5]
Unexposed Age 0.6y	Exposed Age 1y	Unexposed Age 1.3y	Exposed Age 2y	Unexposed Age 0.5y
2.7	3.2	3.7	3.1	1.7

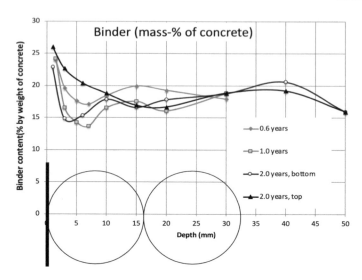

Fig. 3.1 Measured binder content as a function of depth

Fig. 3.2 Liquid saturation as a function of cover depth after 2-year exposure

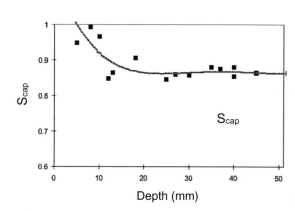

3.2.2 Specimen Dimensions and Environmental Conditions

Specimen were 1 m high, 0.7 m wide, and 0.1 m thick. They were placed vertically, such as having specimen parts exposed to submerged, tidal, and splash zones (see Figs. 3.3 and 3.4). Measurements for submerged case were done at an immersion depth of 0.6 m below seawater level.

The annual average water temperature of the seawater is equal to 11 °C. It varies from 2 to 20 °C depending on the season. Therefore, the surface seawater chloride concentration varies with temperature from 10 g/l to 18 g/l with an average of 14 g/l.

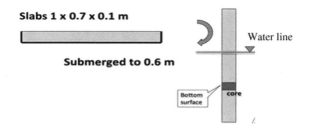

Fig. 3.3 Schematic impression of the slab dimension, its vertical position in the water and the place of coring under water

Fig. 3.4 Impression of the test site and sensor instrumentation

3.2.3   *Experimental Chloride Profiles*

Total chloride profiles were measured after 0.6, 1 and 2 years of exposure. The profiles at 0.6 and 1 year were taken from the bottom surface of a submerged slab. At two years, two profiles were taken, one from the bottom and one from the top of the submerged slab. They are shown in Figs. 3.5 and 3.6. No free chloride profiles were measured.

3.2.4   *Additional Data*

From the chloride profiles the "achieved surface chloride contents" C_{sa} were determined by fitting the error function given by expression 2.4.2. These data are input

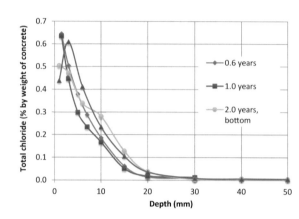

Fig. 3.5 Measured total chloride content as a function of cover depth by % of mass of concrete

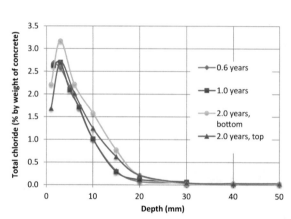

Fig. 3.6 Measured total chloride content as a function of cover depth by % of mass of binder

data for most of analytical models which aim at simulating the total chloride ingress as a function of time. The measured values are exposed in Table 3.3.

Table 3.3 Achieved surface chloride content C_{sa} for concrete from a two-year exposure period

Age (year)	0.63		1.05		2.03				0.6–2
Exposure time (year)	0.6		1.0		2.0 Bottom Top				Average
C_{sa} (wt-% of binder)	3.53*	3.25	3.64*	3.26	3.94*	3.05	3.43*	2.51	3.63
C_{sa} (wt-% of concrete)	0.72		0.70		0.56		0.80*	0.60	

* First depth neglected in the curve-fitting

3.2.5 Benchmark Stages

A first benchmark stage consisted in predicting chloride content (by mass of concrete) profiles after 5, 20 and 100 years of exposure on the basis of the given calibration data (up to 2 years of exposure). Chloride ingress was supposed to evolve in a semi-infinite space. Numerical results were compared to an experimental profile after 5 years of exposure. The scatter between model results were also discussed for 5, 20 and 100 years of exposure.

In a second prediction, modelers were asked to simulate the ingress of chloride from both sides of a 100 mm thick sample. They have been compared to 20 years of exposure experimental data. The main difference compared to previous exercises is that predicted chloride concentrations increased due two-sided ingress. Two concretes (5-40 and 6-40) very similar to the one used in the first prediction exercise were used to compare numerical results to experimental data after 20 years of exposure. The main difference between these new concretes and those from the previous prediction is that in the samples exposed to seawater for 5 years, silica fume was incorporated in the form of slurry, whereas silica fume powder was used in those concretes exposed for 20 years. In addition, less air entraining agent was used (5.9% for the first concrete *vs.* 3% for the second one).

3.3 Analysis and Interpretation

In total, fourteen models have been tested which are detailed in Chap. 2. In general, these models can be subdivided into two categories: analytical (A) and numerical (N) models. In the first (analytical) category, five models are based on the error function, two on Mejlbro-Poulsen model and two on ClinConc. In the second category (numerical), six models are employed: ClinConEng, HETEK-Conv, SDReaM-crete, Double-Multi, Freund/Langmuir and Cerema (Table 3.4).

A first round of benchmark analysis is carried out on all the numerical models calibrated on given experimental results (Tables 3.1 and 3.2 and Figs. 3.5 and 3.6 on chloride profiles after 0.2, 1 and/or 2 years of exposure). Modelers were asked to make predictions for 5, 20 and 100 years of exposure, while experimental data were not shown. This was followed by a second benchmark where modelers were asked to predict chloride ingress in a 100 mm thick sample after 20 years of exposure under the same conditions. In the second modeling round, all the experimental chloride profile results were revealed, and the modelers were asked to make free adjustments of input parameters in order to improve their predictions based on the experimental data for 5 and 20 years of exposure. With this, the present document also reports a more detailed analysis of models, to investigate the role of the modelers and the consequences of their initial input data choices on the final predicted result.

Table 3.4 List of models involved in the benchmark, categorized in Analytical (A) and Numerical (N) models, while also indicated the chloride ingress space condition

Model	Analytical/Numerical	Semi-infinite space, ingress from one side	Finite space, ingress from both sides
fib-Model	A	X	
ERFC-Lafarge	A	X	
ERFC-Cerib	A	X	
ERFC-LASIE	A	X	
ERFC-LNEC	A	X	
Mejlbro-Poulsen	A	X	
HETEK	A	X	
ClinConcEngSimpl	A	X	
ClinConEng	N	X	X
Double-Multi	N	X	X
SDReaM-crete	N	X	X
HETEK-Conv	N	X	X
Freund/Langmuir	N	X	X
Cerema	N	X	X

3.3.1 Global Analysis

3.3.1.1 Chloride Ingress in a Semi-Infinite Space (First Prediction)

Prediction after 5 years of exposure

Figure 3.7 shows the simulated total chloride profiles obtained with all the 14 models after 5 year-exposure. Experimental data are also plotted and show a particular behavior. In the first centimeter of concrete surface layer, chloride profile seems to follow an evolution that is different from the trend observed in deeper inner layers. Another element that can be emphasized is the fact that almost all the experimental data are contained in the envelope consisting in the minimum and the maximum of the simulated results. In other words, the mean of all predictions together is in a quite good agreement with the experimental profile.

However, several profiles are quite different from most others. For example, «Cerema» model predicts a surface content that is drastically lower than the largest (C_s) value. All these observations are discussed in the «specific analysis» part.

In order to deeper analyze the numerical results a root mean squared deviation (RMSD) was calculated. The reference data needed to calculate the RMSD (for each model prediction) was the experimental data themselves or an approximation of the experimental data by fitting the error function as given by expression 2.4.2. In the latter case, the fitted (error-function model) curve was obtained by a nonlinear least squares' method (Levenberg–Marquardt). Assuming that the initial total chloride

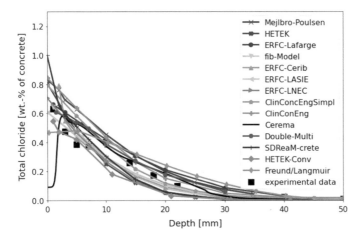

Fig. 3.7 Chloride content prediction after 5 years of exposure

content, C_0, is equal to zero and that the profile follows a diffusion-like behavior described by the error function (*erfc*), estimation of D_{app} and C_s leads to 6.07×10^{-13} m²/s and 6.49×10^{-01} wt.% of concrete, respectively. Figure 3.8 shows the fitted curve obtained and one standard deviation on each side of the curve.

Root mean square deviation (RMSD) calculation based on experimental data.

The RMSD for comparing the modelling results with the experimental measurements are provided in Fig. 3.9. The results are separated into analytical results (upper part) and numerical results (lower part). They clearly show that the deviation in RMSD is larger between analytical models rather than between the numerical ones. They also

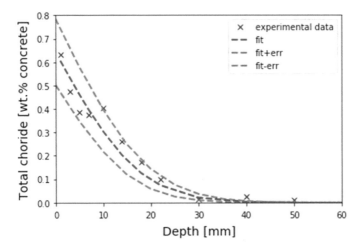

Fig. 3.8 Fitted error function base on experimental data (with RMSD statistical error)

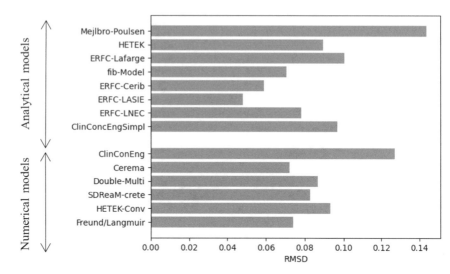

Fig. 3.9 Root mean square deviation (RMSD) between model results and 5-years experimental data

underline the importance of modelers (i.e. human factor), where due to the different choices that can be made (for input parameters) by employing the same model, can yield very different results. This is especially relevant for model results based on the *erfc* function as well as those calculated with the Hetek or ClinConc model. These (first round) results clearly indicate the relevance of an accurate model calibration and understanding the impact of the various model input parameters on the predicted result, already for only 5 years exposure.

Root mean square calculation based on fitted curve through experimental data.

The RMSD results were also determined from the modelling results and the reference values which is considered the best fitted curve through the original experimental data as presented previously (5 years), as well as for the original experimental data. The two RMSD calculations are provided in Fig. 3.10 and show very different results. The difference between the results obtained from the two different RMSD calculations are most pronounced for the analytical models based on the *erfc* function. The differences in RMSD results are much smaller for the numerical models, especially when comparing them with the RMSD results obtained for the analytical models based on *erfc* function. For the numerical models, differences are less pronounced and their predictions generally do not vary (overestimate or underestimate the experimental data) to such an extent as (between) the analytical models.

Although some models based on the error function show a larger scatter (variability), most of them are still close to the experimental data. In general, it can be said that results are a bit better for those models that used the measured diffusion coefficient determined by the immersion test rather than the migration test. The results

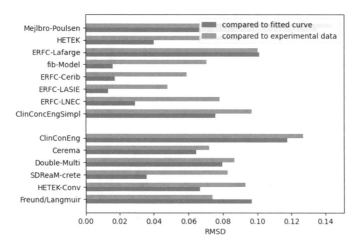

Fig. 3.10 Comparison of root mean square deviation based on 5-years experimental data (orange) and fitted curve (blue)

are discussed in more detail later in Sect. 3.3.2, «specific analysis», where modelling results and differences are motivated explicitly.

Predictions after 20 and 100 years of exposure

Model predictions were also computed for an exposure time of 20 and 100 years, where only for the 20 years exposure condition experimental data was available. For the 100 years exposure time, no experimental data was available, meaning that only the various modelling results are compared with each other. Detailed experimental data for the 20 years exposure condition were available, however, it had been observed that for some of the 100 mm thick test samples, the chloride ingress results had overlapped as they were diffusing into the test samples from both sides. The results achieved from the various models are presented and described in the next section. Figure 3.11 shows the modelling results for 20 years exposure, whereas the 100 years of (predicted) exposure time is presented in Fig. 3.12. The figures show the total chloride profiles computed with both the analytical and numerical models. The figures indicate that the scatter among the various modelling results increases with time of exposure, i.e., 20 years *vs.* 100 *years*. This increase in variability affects both the chloride content along the cover depth, but also for the surface content (C_s).

Apparent diffusion coefficient and surface chloride content

From the various modelling results presented in the figures (Figs. 3.7, 3.11 and 3.12), the evolution of the apparent diffusion coefficient and chloride surface content as a function of time was obtained by employing inverse analysis (back-fitting) method. The information receiving from this calculation is in particularly interesting for the durability assessment of concrete structures. In the scientific community, but also in the daily practice, there is still an ongoing discussion about the impact these factors have on the accuracy of the predicted chloride ingress results, and with this, on the

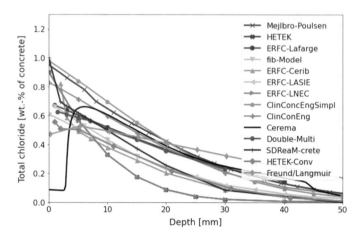

Fig. 3.11 Chloride content prediction after 20 year of exposure

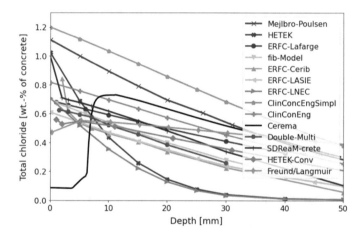

Fig. 3.12 Chloride content prediction after 100 years exposure

reliability of the service life assessment of concrete structures. As a clear physical background of these two input parameters is lacking, they could also be considered as some sort of fit factor that should be calibrated to any concrete structure under consideration. This means that for existing structures, (semi-empirical) evolution of these parameters (with exposure time) may provide reasonable estimates if they can be calibrated with data received from chloride affected concrete structures, at different exposure periods. For new concrete structures, the evolution of these parameters is still under strong discussion as they cannot be calibrated beforehand (depending on initial and boundary conditions), leading to unreliable predictions. In that case, their values should be "assumed" based on expert judgements, or, whenever possible, determined from limited (accelerated) laboratory measurements (i.e.

RCM-test). However, this often leads to very high uncertainties in the (probabilistic) service-life assessment predictions.

From a modeling point of view, both the apparent diffusion coefficient (D_{F2} or D_{av}) and the surface chloride content (C_s), are principal parameters which are more important for analytical model rather than for numerical models, which are instead based on D_{F1} and binding isotherms parameters. Especially for those numerical models which operate with the free chloride concentration, i.e. consider the chemical (non-linear) binding phenomena, the time dependency of C_s and time and space dependency of D_{F2} is (at least partially) implicitly considered. In particular, back-fitting the *erfc* function (solution to Fick's 2nd law at particular conditions) on the predicted profiles (Figs. 3.7, 3.11 and 3.12) provides the ability to obtain the apparent diffusion coefficient as well as the surface chloride content, where the material is considered as a homogeneous material. This also means that D_{av} is supposed not to depend on depth, which is inherent to an analytical model, but could be treated differently in numerical models. Therefore, such back-fit analysis must be considered and interpret with caution. Nevertheless, the two quantities were calculated by inverse analysis for all analytical and numerical models involved to provide a clear tendency on various model behaviors. The results also show how each model treats the time dependency of the apparent diffusion coefficient (see Figs. 3.13 and 3.14) and the expose surface chloride content as a function of exposure time, respectively.

These figures reveal some interesting observations. For example, tor the evolution of the apparent diffusion coefficient (Fig. 3.13), most models predict a decrease of its initial value with time. However, these results are rather obvious for analytical models since this time-dependent reduction in diffusivity is an a priori part of the model hypothesis, and with this, explicitly formulated as a model input. For the numerical models, a different behavior could be observed. For example, the "Freund/Langmuir" model assumes a constant apparent diffusion coefficient which

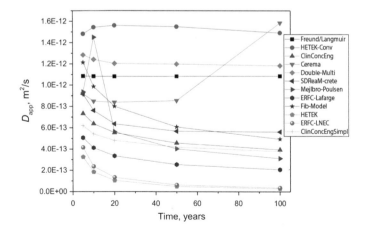

Fig. 3.13 Apparent diffusion coefficient (D_{av}) versus exposure time obtained by inverse analysis (*erfc* back-fitting) on modelling results of profile predictions

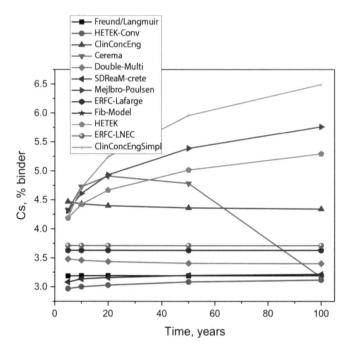

Fig. 3.14 Surface chloride concentration evolution obtained by (*erfc* back-fitting) inverse analysis on modelling results of profile predictions

is the result of a constant D_{F1} diffusion coefficient and a constant dC/dc (i.e. for a linear binding isotherm, $C = k \cdot c$) as a function of time dependent pore solution concentration (c) (see derivation for D_{F2} in Sect. 3.2.2, in particular Eqs. (2.2.4)–(2.2.6). More non-linear description of the chloride binding isotherms, namely by employing different binding input parameter(s), results in (apparent) time dependency of $D_{app,F2}$ (in Sect. 3.2.4, the average of D_{F2} in the time interval $0 - t$ is defined as D_{av}). For the "HETEK-Conv" model, a slight increase is observed in the first 20 years followed by a slow decrease thereafter. For the "Cerema" model, the behavior is completely opposite. The model predicts a slight decrease of the diffusivity up to the first 20 years, followed by a relatively steep increase after 50 years of exposure. These differences in apparent diffusivity are depending on the type of model and the interpretation of the chemical-physical processes that evolve with time. Equation (2.4.3) is a common (very simplified) way to express the time-dependency of the average diffusivity (D_{av}). This assumption should be critically considered and understood. Although the apparent diffusion coefficient might change due to hydration, these changes are negligible after certain time period (typically after 1 year). Moreover, apparent diffusion coefficient is a function of porosity and that porosity may evolve during service life also due to environmentally induced chemical precipitations (e.g. carbonation), cracks, leaching etc., also resulting in the time dependency. However, the primarily effect incorporated in Eq. (2.4.3) is to handle the non-linear effects of the chloride binding. Therefore, this is just a simple assumption made to still be able to use analytical models (based on *erfc* solution). In view of chloride binding, the most critical assumption of Eq. (2.4.3) is in neglecting of spatial variability of D_{av}. This may become important for non-linear dependence of binding with concentration which is not considered by common analytical models. Finally, although field data show promising modeling results when using this equation (Eq. 2.4.3), we still do not fully understand why. Most probably the multi-species approach for more detailed description of the many underlying physical–chemical interactions (and pH dependence) could answer this question, but it requires much more research to independently obtain numerous relevant parameters.

Regarding the evolution of the surface chloride content with time (Fig. 3.14), the predictions reveal even higher variability. For the analytical models that are based on the *erfc* function, the surface chloride content is not supposed to be time dependent, which is also reflected by the results presented in Fig. 3.14. Models based on "Mejlbro-Poulsen" model ("HETEK" and "Mejlbro-Poulsen") show a strong increase of the surface content with time, whereas a much more moderate increase was predicted by some numerical models such as "HETEK-Conv" and "SDReaM-crete" models. The two numerical models ("ClinConcEng" and "Double-Multi" models) show a slight monotone decrease of the surface chloride content, whereas the "Cerema" model shows a strong increase over the first 20 years, followed by a rapid and progressive decrease. For the "HETEK-Conv" model, it should be noted that the surface content results are computed by extrapolating the numerical results to the y-axis. However, for the "HETEK-Conv" and "Cerema" models, the predictions not always show a monotone behavior (for various reasons). Therefore, the calculated surface contents for these models could also be very different from those deduced by extrapolation.

3.3.1.2 Numerical Prediction of a 100 mm Thick Sample (Second Prediction)

Analytical models are derived from Fick's 2nd law or Psi function (for Mejlbro-Poulsen model) and are only applicable for semi-infinite space (i.e. infinitive depth) domains. Because of this, they cannot be used for calculating the chloride ingress of relatively thin samples or structures, exposed to chlorides which penetrate from both sides in opposite direction (resulting in finite depth domain). Reason is that the chlorides diffusing from both sides, may interfere with each other leading to an accumulation over the full depth of the concrete element. For numerical models, this limitation does not exist since for the discretization of the differential equation of Fick's 2nd law, the actual geometrical dimensions and two-sided boundary conditions can be easily considered. In this benchmark case, the data was taken from samples with a thickness of 100 mm, which turned out to be relatively thin, as the ingress fronts met already after 20 years of exposure. Therefore, for 'finite' cases only numerical models were considered where eventually increased chloride ingress content due to overlapping after long time exposure times could be considered explicitly (see Fig. 3.15).

Concerning experimental data, several remarks should be made. Firstly, the maximum total surface chloride content is not measured on the first point close to the surface but on the second point, although samples were completely immersed. Secondly, the total chloride content is almost constant beyond 25-mm depth. This phenomenon is due to the ingress of chloride from both exposure sides caused by the limited sample width of 10 cm, leading to an overlap of the chloride ingress contents. Concerning the prediction accuracy of the calculated chloride ingress profiles, several models clearly overestimated the correct slope indicated by the experimental data. This tendency is also visible in Fig. 3.16, which exposes the envelope (maximum and

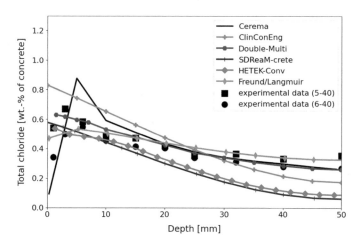

Fig. 3.15 Predictions on 100 mm thick samples with chloride ingress from both sides after 20-years exposure

Fig. 3.16 Envelope of numerical results for chloride ingress from both sides after 20-years exposure

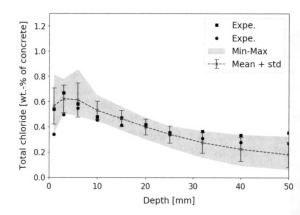

Fig. 3.17 Root mean square deviation based on experimental data after 20-years exposure

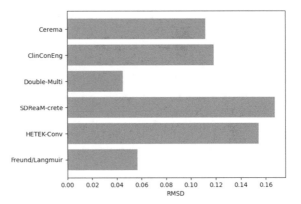

minimum) of the predictions as well as the mean value and the standard deviation of all modelling results employed in Fig. 3.16.

The RMSD for each model result was calculated by considering the experimental data as a reference, and are provided in Fig. 3.17. From these RMSD values it can be observed that the results of "Cerema" and "ClinConcEng" models are very close to each other. Also, the "SDReaM-crete" and "HETEK-Conv" models show almost identical results. However, although these model predictions are almost equivalent, they tend to underestimate the experimental data. On the other hand, the 'ClinConcEng' model prediction overestimates the measured data before 25 mm ingress depth but underestimates it thereafter (see Fig. 3.15). The 'Cerema' model prediction shows a dramatic overestimation before 15 mm ingress and approaches the experimental data quite closely thereafter.

3.3.2 Specific Analysis by Grouping Models

Numerical results can also be compared by grouping various models which have the same basis. Such comparison aims at identifying the underlying choices modelers made for their specific analysis. Four groups have been identified, where the first group brings together models that are based on the analytical *erfc* function, the second group is based on the Mejlbro-Poulsen solution, the third on the Clinconc model, and the last one encompasses the numerical models that need a numerical solution strategy.

3.3.2.1 Simulations Based on *erfc* Function

For this benchmark test, five simulations have been computed on the basis of the *erfc* function. Two main approaches have been considered, where in one approach modelers deduced the input parameters by curve fitting the data from the chloride profiles, while the other modelers choose to use the measured diffusion coefficient and surface content for the benchmark.

Theoretical background

The analytical solution of Fick's second law is as follows:

$$C(x, t) = C_s \cdot \text{erfc}\left(\frac{x}{2\sqrt{D_{av} \cdot t}}\right) \qquad (3.3.1)$$

where t is the time of exposure and D_{av} the mean value of the diffusion coefficient during the exposure time. D_{av} can be written as follows:

$$D_{av} = D_{av,ex} \cdot \left(\frac{t_{ex}}{t}\right)^{\alpha} \qquad (3.3.2)$$

with $D_{av,ex}$ the diffusion coefficient measured at t_{ex} which is normally defined as 28 days. The parameter α is the ageing parameter (also called ageing factor).

In the first approach, only experimental chloride data profiles with a relatively short exposure time (generally up to 2 years) were used to predict the long-time chloride ingress. This methodology is of particular interest whenever no diffusion coefficient was initially measured. It that case the following steps could be considered:

- Estimating the diffusion coefficient and chloride surface content by regression analysis on given experimental profiles by fitting the Eq. (3.3.1), and by applying the least-squares method.
- Estimating the diffusion coefficient D_0 at the first time of exposure to sea water t_0 and the ageing parameter n by regression analysis fitting Eq. (3.3.2), to the diffusion coefficient values estimated in the first step.
- Use the deduced values for D_0, n and C_s in Eqs. (3.3.1) and (3.3.2) to predict the chloride profiles at later ages.

In the second approach, the diffusion coefficient will be determined from experimental measurements (migration or non-steady state diffusion tests), whereas ageing parameter (n) will be based on assumptions that consider the concrete composition and exposure class. For two groups of models, the apparent diffusion coefficient is taken equal to the measured migration diffusion coefficient after 180 days of exposure. For another group, two different coefficient values were determined, i.e., one from a migration test (average value) and another one from an immersion test. In two other cases, the diffusion coefficient (at 28 days) was deduced from the results measured at different ages using Eq. (3.3.1).

Concerning the ageing factor, LafargeHolcim choose 0.3 which correspond to an OPC after fib bulletin ("Model Code for Service Life Design"), Cerib choose 0.45, which is based on a hypothesis for their binder used, and Lasie employed 0.4, which was based on a literature review and an equivalent concrete (95% CEM I + 5% Silica Fume). Temperature, Treal (Kelvin) can be different depending on the modelers.

Table 3.5 gives an overview of input parameters employed by modelers. It may be important to note that all modelers have employed a surface content that is constant with time. Knowing that the *erfc* function is an analytical solution of Fick's second law, selecting a surface content that does not depend on time, seems to be very reasonable. All modelers have also chosen to employ a diffusion coefficient that decreases with time (e.g. Eq. 3.2.2). Generally, such choice is rather questionable. However, for short time simulations such approximation can be adopted with reasonable accuracy. For longer time simulations, the results may drastically underestimate the total chloride content if the diffusion coefficient decreases unlimited.

The chloride ingress results calculated with various models and for different times of exposure are provided in Figs. 3.18, 3.19 and 3.20. From these results, several observations can be underlined and explained in more detail. Firstly, as expected,

Table 3.5 Input data for *erfc* function

	ERFC-LNEC	Cerib	Lafarge D_{app}	Lafarge D_{mig}	Lasie
t_0	14 days	180 days	28 days	28 days	180 days
n	0.823	0.45	0.3	0.3	0.4
T_{real} (K)	293	284	284	284	284
D_m (10^{-12} m^2/s)	22.21	2.7	7.14	2.98	2.7
C_0 (wt-% of binder)	0.03	0.05	–	–	0.05
Δx	0	0	0	0	0
C_{sa} (wt-% of binder)	3.72	3.63	3.63	3.63	3.20

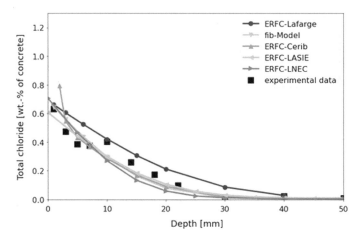

Fig. 3.18 Chloride profiles after 5 years exposure for the *erfc* function-based models

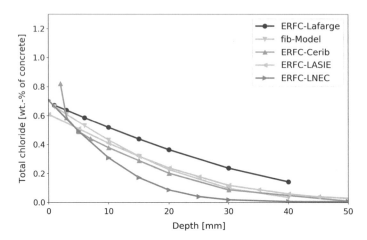

Fig. 3.19 Chloride profiles after 20 years exposure for the *erfc* function-based models

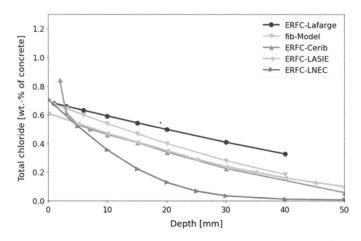

Fig. 3.20 Chloride profiles after 100 years exposure for the *erfc* function-based models

the scatter between the individual modelling results turns out to enhance progressively with time of exposure. Secondly, the predictions conducted with a diffusion coefficient deduced from a migration test results in nearly the same chloride profiles (LafargeHolcim_Dmig, Lasie, Cerib), even for long time exposures. Prediction with a diffusion coefficient deduced from immersion test leads to larger chloride content values. Only the ERFC-LNEC simulations show somehow deviating results in comparison with the other analytical *erfc* models. This is due to the fact that for this model, the ageing factor was estimated from the measured chloride ingress profiles (see Figs. 3.5 and 3.6). A deeper assessment on the particularities/limitations of estimating the ageing factor from the chloride profiles at young ages is still to be considered.

3.3.2.2 Results Based on Numerical Mejlbro-Poulsen Model

Two modelers employed the Mejlbro-Poulsen model to predict the chloride ingress into the concrete cover. Such model is particularly adopted for predicting the total chloride ingress whenever both the diffusion coefficient and the chloride surface content are time dependent. This situation requires an extension of the standard *erfc* model, which is done according to the following theoretical backgrounds:

Theoretical background

As discussed previously, the standard *erfc* function, being the analytical solution of the Fick's second law, is restricted to constant surface contents C_s. In case of time-dependent surface contents, the analytical solution can be written according to the follow formulation:

$$c(x, t) = C_S \Psi_p \left[\frac{x}{\sqrt{4 D_m t}} \right] \tag{3.3.3}$$

where C_s is as follows:

$$C_S = S[D_m.t]^p \tag{3.3.4}$$

where S is the surface chloride content coefficient of concrete and p an exponent that considers the time-dependency of the surface chloride content of concrete. As in the standard *erfc* function model, D_m is obtained from Eq. (3.3.2).

As in the use of the standard *erfc* function-based model, one modeler chose to deduce the model parameters from the experimental profiles only (see Fig. 3.5). In this case the simulation steps were as follows:

- Estimating the diffusion coefficient and chloride surface content by regression analysis on a given experimental profile while employing Eq. (3.3.1), and using the least-squares method on this *erfc* equation for optimizing the fit. The ageing factor was taken the same as used for the Hetek and standard *erfc*-based model.
- Estimation of S and p was done with Eq. (3.3.3), where C_s is estimated in previous steps and considering $D_m = \mu_p D_{m,erfc}$, $D_{m,erfc}$ the diffusion coefficient estimated with *erfc* function-based model and $\mu_p \approx 1 + 0.5194p - 0.0876p^2 - 0.0022p^4$;
- Estimating D_0 by the expression $D_0 = \mu_p D_{0,erfc}$ where $D_{0,erfc}$ is the diffusion coefficient at time t_0 from *erfc* function-based model.
- Using the deduced values of D_0, n, S and p in Eqs. (3.3.3) and (3.3.4).

The values S and p were obtained from an inverse analysis, and it turned out that they are very sensitive to important variabilities, which is the reason that another

Table 3.6 Input data for Hetek based models

		HETEK-LNEC (curve fitting)		Mejlbro-Poulsen
t_0		14 days	D_1	66.8
n		0.823	D_{100}	12.3
D_m (10^{-12} m^2/s)		27.03	C_1	4.0
C_0 (wt-% of binder)		0.03	C_{100}	6.3
S		295.188		
p		0.449		

method was employed by another modeler to determine the parameters. It consists of calculating the diffusion coefficients, D_1 and D_{100}, and surface contents, C_1 and C_{100}, after 1 and 100 years of exposure, by knowing the concrete composition (Frederiksen et al., 1997, pp. 129–131). Table 3.6 shows input data of the modelers.

Similar to the *erfc* function-based model, the use of the diffusion coefficient estimated from the measured chloride profiles in the Mejlbro-Poulsen model, has led to an overestimation of the ageing parameter (n, in Eq. (3.3.2)). This overestimation caused a significant decrease of the diffusion coefficient as a function of time, leading to a strong reduction of the chloride ingress with time. This effect can be seen in Fig. 3.23 where the profiles after 100 year-exposure is relatively close to those at 5 and 20 year-exposure (Fig. 3.23). Moreover, for the two modelling results presented in this section, an increase of surface content can be clearly observed and underlines the model's ability to consider this effect, unlike the analytical models described by the error function (Figs. 3.21 and 3.22).

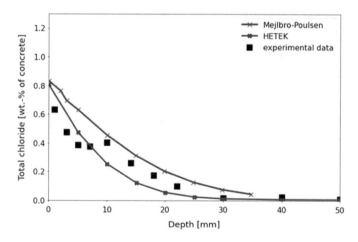

Fig. 3.21 Chloride content profiles after 5-year exposure for the Hetek based models

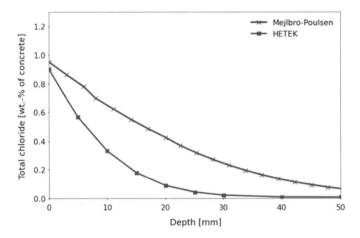

Fig. 3.22   Chloride content profiles after 20-year exposure for the Hetek based models

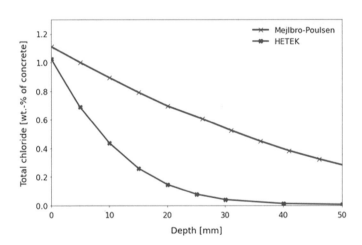

Fig. 3.23   Chloride content profiles after 100-year exposure for the Hetek based models

3.3.2.3   Numerical Simulations Based on ClinConc Model for Engineers

Two modelers used a model based on the ClinConc model. Knowing that the full version of the ClinConc model requires 40 input parameters, one modeler used a simplified version, which was set up with the following 6 parameters:

- Porosity,
- Chloride binding parameters,
- Migration coefficient—determined after 6 month of curing, D_{6m}, preferably in place,
- Surface chloride concentration,

- Internal chloride concentration,
- Temperature.

For the diffusion coefficient the migration coefficient after 6 months curing period, D_{6m}, was adopted. The surface chloride concentration and temperature are given by the sea water concentration and corresponding temperature (Table 3.7). The variation that can arise due to wall effect was not considered as this is a phenomenon that is not uniform across the height of the column/beam. For the benchmark test case, the porosity was not provided. Therefore, the given initial data set (Table 3.7) was employed to determine a fitting parameter for porosity. Chloride binding was assumed to be based on the binder details for a typical CEMI + MS cement.

The other ClinConc modeler, used input data that were basically the concrete mix proportions, chloride migration coefficient, and the exposure conditions (chloride concentration and temperature of the seawater). All other parameters were selected or calculated based on Appendix A in [3].

Table 3.7 Input parameters for ClinConc models

Parameters	ClinConcEngSimpl	ClinConcEng
D_{6m}	2.7×10^{-12} (m^2/s)	2.7×10^{-12} (m^2/s)
C_s	14 (g/l)	14 (g/l)
T	284 K	284 K
Chloride binding parameters	$F_b = 34.65$; $\beta_b = 0.38$; $a_t = 0.33$	$F_b = 34.26$; $\beta_b = 0.38$; $a_t = 0.38$
Initial (internal) chloride content	$C_{ini} = 0.005$ g/l	$C_{ini} = 0.005$ g/l
Porosity	0.11	0.11

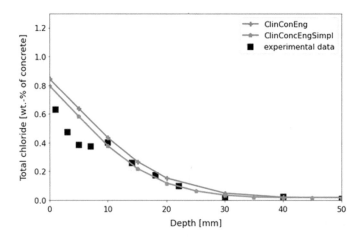

Fig. 3.24 Chloride content profiles after 5-year exposure for the ClinConc based models

Results provided in the figures Figs. 3.24, 3.25 and 3.26 show a very similar behavior for the two predictions employed with the ClinConc models after 5 years of exposure. The numerical results tend to deviate from each other as the exposure period increases to 20 years and even more after 100 years. The main difference is the way how the two predictions consider the evolution of the surface concentration. For the full version of the model a slow decrease of the surface concentration was supposed, whereas a relatively fast increase was assumed in the simplified version. The consequence of this simplification turns out to be significant for the longer-term predictions of the chloride ingress.

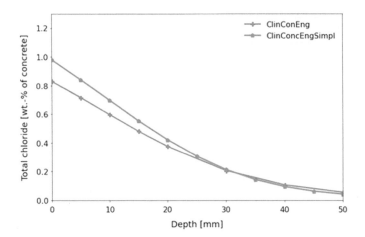

Fig. 3.25 Chloride content profiles after 20-year exposure for the ClinConc based models

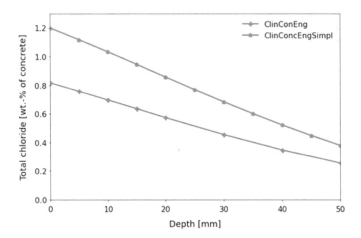

Fig. 3.26 Chloride content profiles after 100-year exposure for the ClinConc based models

3.3.2.4 Other Numerical Models

The specifics of the following numerical models lies in the fact that the computed chloride ingress focuses on free chlorides in particular. It has the advantage that realistic boundary conditions can be applied for submerged cases (chloride concentration is the concentration of the seawater). However, the main problem is the difficulty to find the relationship between free chlorides and total chloride concentration, which is normally expressed in terms of binding isotherm. Several different strategies have been employed in the following models. Two simulations were carried out with the ClinConcEng model. The three simulations are conducted with models that are more specifically adapted for research purposes at that moment.

Input data employed by the modelers

Cerema model

As written in the model description part, the Cerema model needs the following input data:

- Binder composition,
- Concrete mix-design,
- Porosity,
- Water temperature evolution,
- Diffusion coefficient D_{Fl} (effect of pore morphology on diffusion of free ion).

However, only the mix-design and water temperature were provided by the benchmark information.

The model setup and first calibrations were done on cement and silica fume composition (see Table 3.8 Cement and silica fume compositions).

The porosity was assumed to be equal to 12%. The DF1 diffusion coefficient was calibrated on the experimental data after 2-year exposure. Inversion result leads to a value of 0.8 10–12 m2/s. The water temperature was supposed to be constant and equal to 14 °C.

Double-Multi model.

The benchmark concrete modelled in this study was considered to be a composite material with three phases, consisting of cement mortar matrix, aggregates and ITZ. Aggregates were supposed to be polydispersed spheres and ITZs with a thickness of 40 μm were positioned like uniform aureole shells around the aggregate particles. The salinity and water temperature adopted were [Cl−] = 14.2 g/l and T = 15 °C

Table 3.8 Cement and silica fume compositions

	SiO_2	Al_2O_3	Fe_2O_3	CaO	MgO	Na_2O	K_2O	SO_3	CO_2
CEM I	21.39	3.49	4.16	65.12	0.82	0.12	0.3	2.86	0.88
Silica fume	94.75	0.07	0.08	0.34	0.28	0.24	0.7	0.05	–

(from given values [Cl−] = 14 ± 4 g/l, T = 11 ± 9 °C). The diffusion coefficients in the different phases are shown in.

Table 3.9. The effects of other ionic species were ignored owing to the relative low concentrations in the concrete specimen and the relatively small effect of ionic interactions considered. The ionic transport parameters employed in this numerical study are listed in.

Table 3.9, and the initial and boundary conditions are shown in Table 3.10.

Cerib model.

For the Cerib model, the quantity of calcium in each hydrate is needed and was calculated along with some data provided by the benchmark document:

- Quantity of cement (399.0 kg/m^3) and additions (silica fume in this case, 21.0 kg/m^3),
- Type of cement (CEM I) and the quantity of CaO in the binder (61.7%),
- Volume of the cement paste (30.4%),
- w/b ratio (0.4).

The porosity of the cement paste was calculated from the Power's model [Taylor04]:

$$\phi_{cement\ paste} = \frac{\frac{W_{eff}}{B}}{\frac{W_{eff}}{B} + 0.32} - 0.53 * \alpha * \left(1 - \frac{\frac{W_{eff}}{B}}{\frac{W_{eff}}{B} + 0.32} \right)$$

where W_{eff}/B is the effective water on the binder content ratio. α is the hydration degree taken as:

$$\alpha = 1 - e^{\left(-3.3 * \frac{W_{eff}}{B} \right)}$$

Table 3.9 Chloride transport properties in different phases

Field variable	Chloride ions
Diffusion coefficient in aggregates, DA	0
Diffusion coefficient in bulk mortar, D1 × 10^{-12} m^2/s	1.80
Diffusion coefficient in ITZ, D2 × 10^{-12} m^2/s	5.40

Table 3.10 Initial boundary conditions of individual species

Field variable		Chloride ions
Concentration boundary conditions, g/l	x = 0	14.2
	x = 1	0
Initial conditions, g/l	t = 0	0

The concrete porosity is estimated as a weighted average between the porosity of the aggregates and the porosity of the cement paste. As this information was not provided by the benchmark information, an aggregate quantity of 1830 kg/m³ and an aggregate porosity of 3.44% were assumed. The calculated porosity of the concrete was 14.2%.

HETEK-Conv model.

Predictions have also been made with the HETEK-Conv Model. This model is based on the following characteristics:

- Chloride diffusion described by Fick's 1st law, with the concentration of free chlorides as the transport potential.
- A moisture dependent chloride diffusion coefficient.
- A moisture dependent, non-linear chloride-binding isotherm.
- One part of the moisture transport, not carrying ions, described with the water vapor concentration as the transport potential. The moisture transport coefficient δ is assumed to be a constant.
- Another part of the moisture transport, carrying ions, described with the "equivalent" relative humidity RH_{eq} as the transport potential. The liquid transport coefficient k_{RH} is moisture dependent.
- A moisture sorption isotherm that is dependent on the concentration of free chlorides in the pore water.
- The boundary conditions can be time dependent.

Freund/Langmuir model.

In this model, the free chloride ingress is supposed to occur by diffusion only in a fully saturated material. Porosity is deduced from the water to cement ratio using Power's model. A Freundlich isotherm, $C_s = aC^b$ (or Langmuir) is used to describe relationship between bound, C_S, and free chloride, C. First, (Freundlich) parameters a and b are calculated by inverse analysis on the given chloride profiles after 6 months, 1 and 2 years of exposure. The mean values are then used in order to obtain the D_{F1} diffusion coefficient value from a second inverse analysis.

In another approach, the two parameters which were calibrated by Levenberg–Marquardt optimization (in Matlab, and using *pdepe* numerical solver (Ukrainczyk & Koenders, 2016)) are diffusion coefficient (D_{F1}) and one of the (Langmuir) binding parameter, while the binding capacity was obtained from total surface concentration and boundary condition for free concentration (inspired by Baroghel-Bouny et al. CCR, 2012).

Benchmarking Results

Individual predicted chloride ingress profiles as well as envelopes, root mean squared deviations and standard deviations for the various numerical models are provided in the Figs. 3.27, 3.28, 3.29, 3.30 and 3.31 For all predictions with the different models, the scatter between results increases with time of exposure. The first impression that emerges is the fact that profiles are not as regular and monotone as those simulated

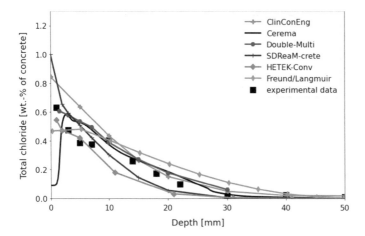

Fig. 3.27 Chloride content profiles after 5-year exposure for numerical models

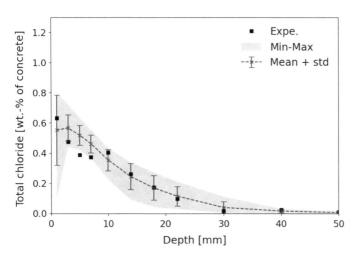

Fig. 3.28 Envelope of numerical results for 5-year exposure

with the analytical models. This is evident for the Cerema model, which considers the mineral dissolution and precipitation explicitly. This is also visible for the HETEK-Conv model, which considers the moisture transport explicitly, and the SDReaM-crete model, that even computes the carbonation reactions explicitly. For these last two models, such a phenomenon is considered because of its relevance at the surface region. For the Double-Multi model, profiles are not regular because of the effect of interfacial transition zone that is modeled explicitly (Fig. 3.32).

Concerning the root mean squared deviations observed after 5 years of exposure, the numerical model predictions are very close to each other. Such a tendency is also confirmed by the envelopes at 5, 20 and 100 years of exposure, where the scatter is

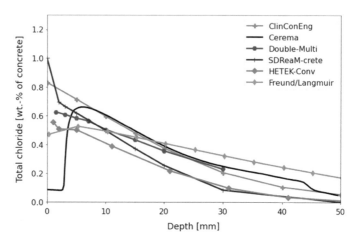

Fig. 3.29 Chloride content profiles after 20-year exposure for numerical models

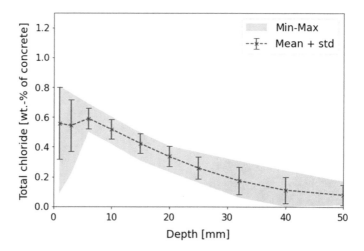

Fig. 3.30 Envelope of numerical results for 20-year exposure

relatively small, in particular after 10 to 15 mm of ingress. This result is particular encouraging given the average concrete cover depth on the rebar is larger than 30 mm.

3.4 Conclusions

From this marine submerged modeling benchmark case, several conclusions can be drawn:

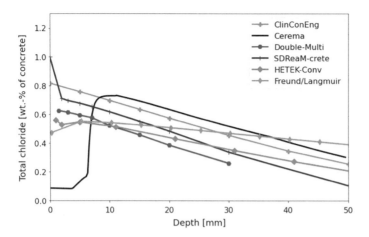

Fig. 3.31 Chloride content profiles after 100-year exposure for numerical models

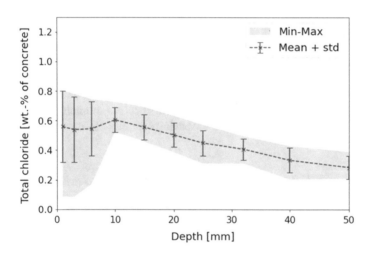

Fig. 3.32 Envelope of numerical results for 100-year exposure

- Model predictions for chloride ingress of concrete submerged in seawater for 5 years show very similar results (Figs. 3.7, 3.8 and 3.9), whatever the model chosen.
- Scatters between model predictions increase with time up to 100 years of exposure at a depth of 5 mm (Figs. 3.12, 3.20, 3.33) reach a maximum standard deviation of 0.112 (for a mean value 0.263).
- After 100 years of exposure, at 5 cm depth, a standard deviation of 0.146 was calculated for the analytical models and 0.0627 for the numerical ones.

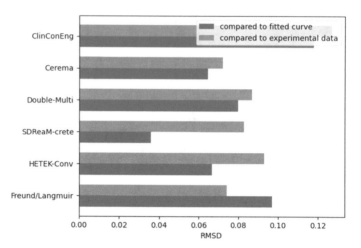

Fig. 3.33 Root mean square deviation for numerical models based on experimental data after 5-year exposure

- For 20 years chloride ingress from both side of a sample, only numerical models are able to achieve the appropriate simulation results (Figs. 3.15, 3.16 and 3.17); analytical models are limited by their semi-infinite assumption (Fig. 3.11).
- Analyzing results within one characteristic group of models underlined the consequences for the choice of the input parameters selected ("human effect") leading to very different results in some cases, especially for models based on *erfc* function.

References

Baroghel-Bouny, V., et al. (2012). Prediction of chloride binding isotherms of cementitious materials by analytical model or numerical inverse analysis. *Cement and Concrete Research, 42*, 1207–1224.

Frederiksen, J. M., Mejer, J., Nilsson, L.-O., Poulsen, E., Sandberg, P., Tang, L., & Andersen, A. (1997). *HETEK, A system for estimation of chloride ingress into concrete, Theoretical background*. The Danish Road Directorate. Report No.83. https://www.google.com/url?sa=t&rct=j&q=&esrc=s&source=web&cd=&ved=2ahUKEwiF_r6L9Yb2AhWihf0HHRF3BsUQFnoECAYQAQ&url=https%3A%2F%2Fwww.teknologisk.dk%2F_%2Fmedia%2F17047_Hetek%252C%2520Report%2520No%252083%252C%2520p1%252C%25201997.pdf&usg=AOvVaw3joU_v6YVZfKh7RhgMlrvf

Nordtest. (1995). *Concrete, hardened: Accelerated chloride penetration*. NTBuild 443. Esbo, Finland. http://nordtest.info/images/documents/nt-methods/building/NT%20build%20443_Concrete,%20hardened_Accelerated%20chloride%20penetration_Nordtest%20Method.pdf

Nordtest. (1999). *Concrete, mortar and cement-based repair materials: Chloride migration coefficient from non-steady state migration experiments*.

Tang, L. (1997). *Chloride penetration profiles and diffusivity in concrete under different exposure conditions*. Report E-97:3, Publication - Department of Building Technology, Building Materials, Chalmers University of Technology. https://research.chalmers.se/en/publication/23503

Tang, L., Nilsson, L.-O., & Basheer, P. A. M. (2011). *Resistance of concrete to chloride ingress: testing and modelling*. Taylor&Francis, Spon Press.
Ukrainczyk, N., & Koenders, E. (2016). Numerical model for chloride ingress in cement based materials: Method of lines implementation for solving coupled multi-species diffusion with binding. *Computations and Materials in Civil Engineering, 1*(3), 109–119.

Chapter 4
Road Sprayed

Jan Bisschop and Lars-Olof Nilsson

Abstract In this model benchmarking study the results of the Bonaduz field experiment in Switzerland, representing a road environment with sprayed deicing salts, were used. In this field experiment chloride profiles have been measured for over a time span of ca. 20 years, and concrete properties as well as weather data are well documented. This chapter includes a detailed description of the field experiments and the simulation/prediction results of 5 participating models. The benchmarking results illustrate the complexity of making chloride ingress predictions for the road sprayed case. The main reason for the differences among model predictions and differences between predictions and measurements could be the differences in the assumed driving force for chloride ingress.

4.1 Description of Benchmark Field Experiment

4.1.1 Introduction

Benchmarking of chloride ingress models for road environments with de-icing salts (Road sprayed) is a challenging task. Compared to the marine case there are additional difficulties: Firstly, the chloride ingress models need to consider the changing and partly unpredictable boundary conditions in road environments. The amount of chloride reaching the concrete surface needs to be predicted or assumed. The fluctuating climatic conditions affect the moisture state of the concrete, and, along with long-term carbonation of concrete, the chloride ingress rate. These effects need to be considered in the modelling as well. Secondly, the benchmarking relies on data (e.g., chloride profiles) from existing long-term field experiments (10, 20 years), and not

J. Bisschop
TFB AG—Technology and Research for Concrete Structures, Wildegg, Switzerland

L.-O. Nilsson (✉)
Lund University, 22100 Lund, Sweden
e-mail: lars-olof.nilsson@byggtek.lth.se

© RILEM 2022   59
E. Koenders et al. (eds.), *Benchmarking Chloride Ingress Models on Real-life Case Studies—Marine Submerged and Road Sprayed Concrete Structures*,
RILEM State-of-the-Art Reports 37, https://doi.org/10.1007/978-3-030-96422-1_4

many of them exist around the world. Field experiments might initially not have been set-up with the purpose of model benchmarking and therefore show deficiencies in this respect: Certain model input parameters may have been not recorded, and the setting of the field experiment may be specific and not represent a common road environment.

Three road-side field experiments have been considered for the benchmarking in this RILEM-TC. These are Bonaduz in Switzerland (duration of 22 years), Naxberg in Switzerland (duration of 18 years), and Borås in Sweden (duration of 20 years). The Naxberg experiment is located in a tunnel within the Swiss Alps and experiences a high chloride ingress rates as was shown in a previous benchmarking study (Bisschop et al., 2016). The setting of the field experiment (a tunnel), the concrete types (relatively porous), and plate thickness of 10 cm (no half-space approximation) make the Naxberg experiment less suitable for an international benchmarking study. The Borås and Bonaduz experiments are probably comparable types of field experiments. The Borås experiment has been published in various reports and was also used in previous benchmarking studies (Tang, 2005; Tang & Utgenannt, 2007; Tang et al., 2010). In the Borås experiment concrete cubes with a height of 30 cm are situated along a highway. In the Bonaduz experiment concrete plates with a height of 2 m are studied. For the Road sprayed benchmarking exercise in the RILEM-TC 270-CIM, the data (i.e., chloride profiles) from the Bonaduz experiment was used, since it was monitored for the longest period. What follows is a detailed description of this field experiment (Sect. 4.1) and a description of the benchmarking results (Sects. 4.2 and 4.3).

4.1.2   The Bonaduz Field Experiment (Concrete Type and Properties)

This Bonaduz field experiment started in 1995 by the SIKA company with the objective to determine the performance of a corrosion inhibitor. The experiment is under the supervision of the Swiss Society for Corrosion Protection (SGK) and is currently (i.e., in 2019) still running. In the experiment, the long-term chloride ingress and the rebar corrosion progress is being monitored. The SIKA company has kindly given permission to use the chloride ingress data for this and future benchmarking studies. The chloride profiles have been published in yearly reports by the SGK. These reports are not public, but all chloride profiles and other relevant data are reproduced here. The main results of the corrosion monitoring and performance of the corrosion inhibitor have been published (Angst et al., 2015).

The experiment consists of three L-shaped concrete elements that were cast under controlled conditions in September 1995, and placed along the 2-lane highway (N12) near Bonaduz in Graubünden in the same autumn. This highway is part of the Alpine crossing "San Bernardino" which is passed by almost 200,000 heavy trucks each year in addition to personal cars (Angst et al., 2015). The elements have a height of 2 m

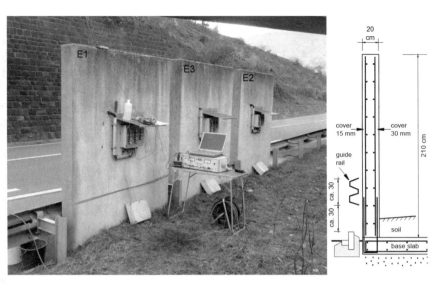

Fig. 4.1 The Bonaduz experiment along the N12-road (photo from internal SGK-report)

and are aligned parallel to road behind a guiding rail (see Fig. 4.1). The experiment is located adjacent an emergency lane and distance from the plates to the main traffic lane is about 2.5 m.

In the field experiment 3 types of concretes are investigated: (i) a reference concrete without inhibitor and without silica fume (concrete E1); (ii) a concrete with a corrosion inhibitor but without SF (concrete E2); (iii) a concrete without inhibitor but containing silica fume (concrete E3). For the benchmarking exercise only the data of the concretes without inhibitor (E1 and E3) have been used. The mixture compositions are given in Table 4.1.

The development of the compressive strength of the concretes is given in Table 4.2. Also included in this table are a few capillary porosity measurements. These properties were measured on laboratory-stored test specimens that were cast from

Table 4.1 Composition of the selected test concretes

Name	w/b	Cement [kg/m³]	SF [kg/m³]	Water [kg/m³]	Paste (CEM I + SF) [vol-%]	Cement content [weight-%]	Calcul. density [kg/m³]	Air content [vol-%]
E1	0.46	325	–	150	25.4	12.5	2526	4.1
E3	0.48	325	20	166	27.7	12.5	2562	3.4

D_{max} = 32 mm (casting date 20.09.1995)
Cement = CEM I 42.5 N; Silica fume = Microsilica (SikaFume-HR) = 6.2% by cement weight; Air-entraining agent (Sika Fro-V5-A) = 0.6% by cement weight; Superplasticizer (Sikament-10) = 0.8% by cement weight

Table 4.2 Concrete properties as function of age (from casting date 20.09.1995)

Property	7 days	28 days	90 days	360 days
Compressive strength[a] [N/mm^2]—**E1**	33	40	47	54
Compressive strength[a] [N/mm^2]—**E3**	34	45	49	54
Capillary porosity [vol-%]—**E1**	–	9.8	10.0	–
Capillary porosity [vol-%]—**E3**	–	10.2	10.5	–

[a]Compressive strength from 15 cm^3 × 15 cm^3 × 15 cm^3 cubes

the same large concrete batch as was used for the test plates along the highway. No data about the moisture content or carbonation progress are available for these concretes. The measured chloride profiles are given in Sect. 4.1.4. The migration coefficients (M_{28d}) were estimated to be 13.3 and 12·10^{-12} m^2/s for concretes E1 and E3, respectively (Bisschop et al., 2016).

4.1.3   Weather and Salt Exposure Data

The Bonaduz field experiment is located at an altitude of 660 m above sea level in the canton of Graubunden. This location experiences pre-alpine winter conditions with moderate snow fall. The annual average temperature is ca. 8 °C. Detailed weather data for the experiment was obtained from the Swiss weather station near Chur (coordinates = 46° 52.2′ N 9° 31.8′ O), located at a 10 km distance from the field experiment. The weather station is 550 m above sea level. The weather measurements can be ordered for the period 2008–2018 from the following website: https://shop. meteoswiss.ch/productView.html?type=psc&id=17. Here, the temperature, relative humidity and snow fall for 2008–2012 are presented (see Figs. 4.2, 4.3 and 4.4), but other weather parameters are available.

The average temperature and RH are plotted in Fig. 4.2 and they show a constant pattern that can be considered to be representative for the whole duration of the experiment. Note that the yearly fluctuation in temperature is in agreement with the assumption in the model proposed by Lars-Olof Nilsson (Nilsson, 2000) for the Swedish situation, with the highest temperature in July/August. However, at Bonaduz, the relative humidity peaks are in December, and not in October as in Nilsson's model (for the Swedish situation). Thus, simplified trends in temperature and relative humidity might be site specific.

In Fig. 4.3 the monthly rainfall is given for the years 2008–2012. The average trend in rainfall shows a peak for the months July and August. The winter months January to March tend to be the driest months of the year. The average monthly trend in rainfall is also plotted in Fig. 4.2 and a rough correlation between temperature and rain fall can be observed. The monthly rainfall trends are relevant to chloride ingress in the Bonaduz experiment since the field experiment is directly exposed to rain. Rainwater flowing down the test plates will affect the moisture state of the concrete

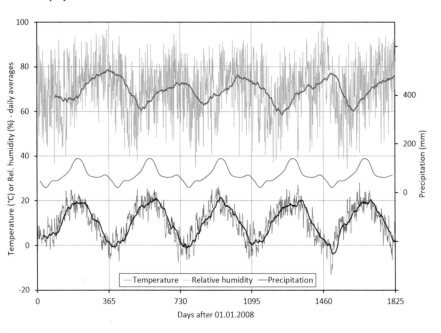

Fig. 4.2 Temperature and relative humidity (daily averages) in the Bonaduz region from 2008 to 2012 (weather station Chur, GR). Trend lines are obtained by applying a moving average. The smoothened trends in rainfall (cumulative monthly amounts) are from Fig. 4.3

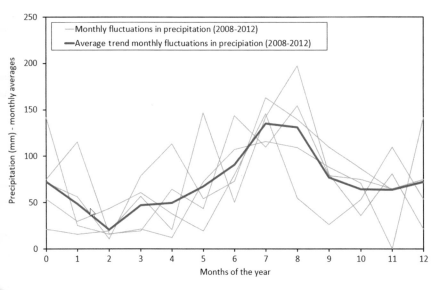

Fig. 4.3 Trends and fluctuations in the monthly rainfall in the Bonaduz region from 2008 to 2012 (weather station Chur, GR). The red line shows the average precipitation trend for the period 2008–2012. This line is repeated 5 times in Fig. 4.2

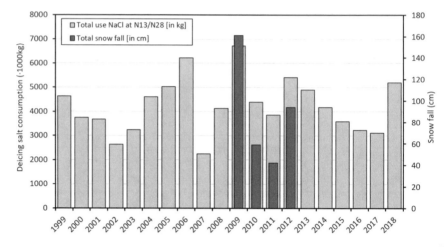

Fig. 4.4 Amount of salt sprayed on roads and snow fall for the Bonaduz region. Salt consumption is total winter season usage for the national roads N12 (Bad Ragaz—Roveredo) and N28 (Landquart—Klosters Selfranga). Winter 2009 means the 'winter season' from 01/07/2008 to 31/06/2009. The snow fall is the cumulative snow fall per 'winter season' from 01/07/2008 to 31/06/2009 for the weather station Chur (GR). *Sources* Oliver Radecke/Tiefbauamt Graubünden and Meteo Schweiz

surface layer. Water ingress may accelerate chloride ingress in wintertime, but it may also lead to leaching of chlorides from the concrete in summertime.

Trends in the deicing salt consumption in the Bonaduz region can be obtained from Fig. 4.4 which shows the total deicing salt (NaCl) usage from 1999 to 2018 at two national roads in eastern Switzerland, including the highway stretch of the Bonaduz experiments: The N12 (Bad Ragaz—Roveredo) and the N28 (Landquart—Klosters Selfranga). These data can be recalculated into salt application amount in kg/m²/winter. That requires determining the total road distance and assuming an average road width. Also plotted in Fig. 4.4 are the snow fall data for the winter seasons 2009–2012. A correlation between snowfall and deicing salt usage can be observed.

4.1.4 Chloride Profiles

Chloride profiles were measured after chloride exposure times of 1.6; 3.9; 4.4; 16; 17.9 and 21.8 years (see Table 4.3). The concrete is exposed to deicing salt at an assumed age of ca. 72 days. The concrete test walls were cast on 20.09.1995. The walls were installed along the highway in autumn of 1995. First assumed chloride exposure is start of winter 1995 (= 01.12.1995). The chloride profiles for concretes E1 und E3 are plotted in Fig. 4.5 and tabulated in Tables 4.4 and 4.5. Chloride profiles were measured below the guiding rail at a height from road of ca. 20 cm, and above the guiding rail at an average height of ca. 100 cm (range: 60–150 cm).

Table 4.3 Chloride exposure durations from assumed exposure start (01.12.1995)	Sampling date Cl-profiles	Years	Days
	01.12.1995	0	0
	01.07.1997	1.6	577
	05.11.1999	3.9	1434
	14.04.2000	4.4	1594
	25.11.2011	16	5835
	05.11.2013	17.9	6545
	12.09.2017	21.8	7951

The samples for chloride content measurement were obtained by drilling into the elements. At each sampling location, 5–6 holes were drilled within an area of ca. 10 cm × 10 cm with a drill of diameter 10 mm leading to a sampling volume at each depth interval, which is in agreement with European standard EN 14629 (Angst et al., 2015). The chloride content in the concrete powder was analyzed according to SN EN 14629. Note that the sampling depth intervals were either 5 mm or 10 mm, and the measurement points are therefore at depths of 2.5, 7.5, 12.5 cm, etc. or at 5, 15, 25 cm, etc. The sampling intervals of 10 mm are indicated in Tables 4.4 and 4.5. The chloride profiles were not all sampled in the same month of the year, and some seasonal fluctuations in the chloride content of the concrete may have been recorded by the data. The chloride profiles measured after 3.9, 16, 17.9 and 21.8 year were all measured in autumn (September–November), and perhaps show only a limited scatter due to seasonal fluctuations in chloride content.

The following trends in chloride profiles can be observed:

- No significant (long-term) effect of height from road on chloride ingress rate in the range of ca. 20–100 cm has been recorded in the Bonaduz field experiment.
- After 16 years of deicing salt exposure there was little change in the chloride profiles for the reference concrete E1. A stagnation of the net chloride ingress is suggested by these data. For the concrete with silica fume (E3) an ongoing chloride ingress (up to 22 years) has been recorded, most strongly in the depth levels down to 30 mm.
- There is no clear effect of silica fume on chloride ingress in the Bonaduz experiment. Perhaps an initial ingress delay can be observed from the data for the concrete with silica fume. However, after 22 years of chloride ingress, the chloride profiles for E1 and E3, especially at depth level of 30–50 mm (= chloride penetration front) are comparable.

Fig. 4.5 Chloride profiles
from the Bonaduz
experiment for concrete E1
and E3

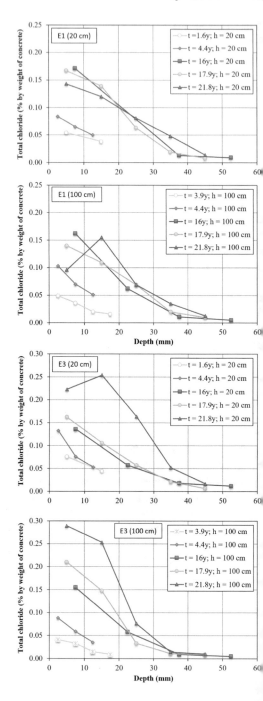

Table 4.4 Bonaduz chloride profiles for concrete E1 at heights of 20 cm (left) und 100 cm (right)

Depth (mm)	Chloride exposure years:					Depth (mm)	Chloride exposure years:				
	1.6	4.4	16	17.9	21.8		3.9	4.4	16	17.9	21.8
2.5		0.084				2.5	0.049	0.103			
5	0.054			0.167	0.143	5				0.140	0.096
7.5		0.065	0.171			7.5	0.036	0.070	0.162		
12.5		0.050				12.5	0.020	0.051			
15	0.038			0.139	0.120	15				0.109	0.155
17.5						17.5	0.016				
22.5						22.5			0.062		
25				0.063	0.081	25				0.069	0.069
35				0.019	0.048	35				0.020	0.035
37.5			0.013			37.5			0.011		
45				0.008	0.014	45				0.010	0.013
52.5			0.009			52.5			0.005		

Chloride content is expressed as a percentage of weight of (dried) concrete

Table 4.5 Bonaduz chloride profiles for concrete E3 at heights of 20 cm (left) und 100 cm (right)

Depth (mm)	Chloride exposure years					Depth (mm)	Chloride exposure years				
	1.6	4.4	16	17.9	21.8		3.9	4.4	16	17.9	21.8
2.5		0.133				2.5	0.041	0.088			
5	0.076			0.162	0.223	5				0.210	0.289
7.5		0.076	0.136			7.5	0.033	0.059	0.155		
12.5		0.053				12.5	0.015	0.035			
15	0.044			0.106	0.254	15				0.148	0.253
17.5						17.5	0.009				
22.5			0.057			22.5			0.058		
25				0.057	0.163	25				0.033	0.076
35				0.021	0.052	35				0.011	0.014
37.5			0.018			37.5			0.009		
45				0.007	0.017	45				0.009	0.010
52.5			0.012			52.5			0.005		

Chloride content is expressed as a percentage of weight of (dried) concrete

4.2 Model Predictions for a Road Environment with Deicing Salt

4.2.1 Introduction

In the following sections, the results of the model predictions of the Bonaduz field experiments are presented. Five models participated in the road environment benchmarking exercise: (1) the Mejlbro-Poulsen model (Jens Mejer Frederiksen); (2) the ClinConcOrig-model (Luping Tang); (3) the Cerema-model (Anthony Soive); (4) the Double-Multi-model (Qing-feng Liu); and (5) the HETEK Convection model (Lars-Olof Nilsson). Setting up a well-defined benchmarking exercise for the road environment case turned out to be complicated, and therefore no detailed guidelines for carrying out the predictions were prescribed. The presented predictions intend to illustrate the complexity of predicting chloride penetration into roadside concrete structures. The origin of the predictions differences of the participating models was not identified.

Table 4.6 gives a rough overview of the prediction differences of four participating models. It shows which of the following parameters were considered in the predictions:

A   = seasonal/fluctuating weather conditions (e.g., temperature, humidity, rain, snow, wind);
B   = drying and moisture state of concrete;
C   = carbonation of concrete;
D   = chloride binding;
E   = time-evolution of chloride surface concentration;
F   = concrete properties depending on time (age) or moisture state;
G   = dependence of height from road.

More detailed descriptions of the used model input parameters are given in the following sections and in Chap. 2 and literature references therein.

Table 4.6 Overview of used input parameters/simulation characteristics

Model	Parameter considered in predictions. See text for explanation						
	A	B	C	D	E	F	G
ClinConcOrig	Yes	Yes	No	Yes	Yes	Yes	No
Cerema	Yes	Yes	Yes	Yes	Yes	No	No
Double-Multi	Yes	No	No	Yes	No	No	Yes
HETEK-Conv	Yes	Yes	No	Yes	Yes	No	Yes

Table 4.7 Input parameters
for the Mejlbro-Poulsen
simulations

Concrete constituents etc	E1	E3	Unit
Cement content	100	94.2	%
Micro silica content	0	5.8	%
Effective water content	46	48	%
Air content in concrete	4.1	3.4	% vol
Initial chloride content	0.08	0.08	% mass B
Concrete age at Cl⁻-exp	0.20	0.20	age < ½ year
w/b ratio	0.46	0.48	–
Diffusion coefficients	ATM/DRS	WRS	Unit
D_1	41	22	mm²/year
a	0.33	0.33	–
D_{100}	9	5	mm²/year
Surface concentrations	ATM/DRS	WRS	Unit
C_1	1.55	1.76	% mass B
C_{100}	2.32	2.65	% mass B

4.2.2 Mejlbro-Poulsen Model

The chloride ingress predictions for the Bonaduz field experiment using the Mejlbro-Poulsen model were carried out by Jens Mejer Frederiksen. The Mejlbro-Poulsen model is described in Chap. 2 (Sect. 2.6) and more details can be found in published literature. The input parameters for the simulations of test elements E1 and E3 are given in Table 4.7. The predicted chloride ingresses after 15.6, 17.6 and 100 years are shown together with the measured chloride profiles in Fig. 4.6a, b.

4.2.3 ClinConc-Model

The chloride ingress predictions for the Bonaduz field experiment using the ClinConc-model were carried out by Luping Tang. In the RISE Report 2018-66 "Chloride Ingress and Reinforcement Corrosion" (Tang et al., 2018) the previous ClinConc model for marine environment has been further modified for wet-dry environment such as road or marine splash zone. The basic transport equations are unchanged, but two equations were introduced, one for the peak depth, x_{peak}, representing the maximum chloride concentration in the cover zone affected by carbonation, and another for the redistribution factor, f_{red}, of chloride binding in the surface zone:

$$x_{peak} = \alpha_p \ln\left(\frac{w}{b}\right) + \beta_p \qquad (4.2.1)$$

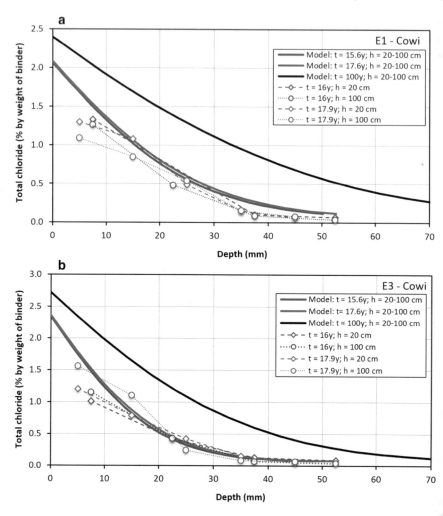

Fig. 4.6 a Modelling results of the Mejlbro-Poulsen model (COWI) and comparison to measured chlorid profiles in the Bonaduz field experiments E1. The model results are independent of the height from road (20 or 100 cm). **b** Modelling results of the Mejlbro-Poulsen model (COWI) and comparison to measured chlorid profiles in the Bonaduz field experiments E3. The model results are independent of the height from road (20 or 100 cm)

where w/b is the water-binder ratio, constant α_p and β_p are empirical constants. For the most types of binder, $\alpha_p = 16.2$ and $\beta_p = 21$. These values were used in this simulation.

$$f_{\mathrm{red}} = \begin{cases} \left[1 - k_1 \cdot c_{\mathrm{f}} \cdot \frac{(x_{\mathrm{peak}} - x)}{x_{\mathrm{peak}}} \cdot \sqrt{\frac{w}{b} - b_0}\right] \cdot k_2 \cdot \sqrt{\frac{w}{b} - b_0} & x \leq x_{\mathrm{peak}} \\ k_2 \cdot \sqrt{\frac{w}{b} - b_0} & x > x_{\mathrm{peak}} \end{cases} \quad (4.2.2)$$

where f_{red} is a redistribution factor to the bound chloride concentrations in the previous ClinConc model, c_f is the free chloride concentration estimated by the ClinConc model, k_1, k_2 and b_0 are constants.

The predicted chloride ingress after 15.6, 17.6 and 100 years are shown together with the measured chloride profiles in Fig. 4.7. For concrete element E1 a larger value of the "drying" input parameter n_d had to be used in order to fit the measured chloride profiles (see Table 4.8). It is this drying effect which makes the profile after 100 years for E1 less progressed as for E3.

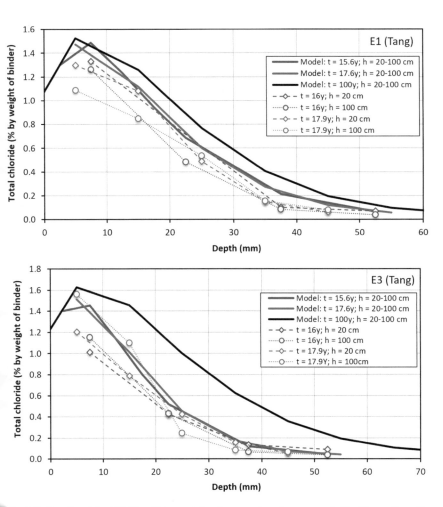

Fig. 4.7 Results of the ClinConcOrig-model and comparison to measured chloride profiles. The model results are independent of the height from road (20 or 100 cm). Model input parameters in Table 4.8

Table 4.8 Input parameters of the ClinConc-simulations for the Bonaduz field experiment

Binder	D_{RCM} $\times 10^{-12}$ m^2/s	c_f g/L	n_d[1]	k_1	k_2	b_0
E1, 20–100 cm	11.2[2]	4	0.8	0.1	1	0.3
E3, 20–100 cm	7.7[2]	4	0.3	0.1	1	0.35[3]

[1] An additional age factor n_d is introduced in the diffusion equation (due to desiccation under the wet-dry environment)
[2] Estimated based on the previous experience for Swedish concrete
[3] Value for concrete with 6% silica fume, similar to that in RISE Report 2018-66

4.2.4 Cerema Model

The chloride ingress predictions for the Bonaduz field experiment using the Cerema-model were carried out by Anthony Soive. For the Cerema-model, cement (or binder) composition, concrete mix-design, porosity and the D_{F1} diffusion coefficient are necessary as input data. The composition of the cement and silica fume were assumed to be as shown in Table 4.9. For the concrete without silica fume (E1), the porosity is assumed to be equal to 10%. Effective diffusion coefficient has been fitted thanks to the first chloride profile (after 1.6 year-exposure) and its value is equal to 1.1e−12 m^2/s.

Boundary condition for the submerged case in seawater (saturated condition and ionic concentration of seawater) are relatively well defined. In the road environment case, the definition of boundary condition is much more complex. Several assumptions had to be made: The first one is the chloride concentration value upon the moment the concrete is exposed to deicing salt. It was supposed to be equal to 1 mol/l. The second one is related to the exposure cycles for which the measured weather data was used. Concrete was assumed to be exposed to deicing salts when snow is present or when it is raining and temperature is below zero. Concrete is exposed to rainwater when it is raining and temperature is above zero. Ionic concentration of deicing salt and rainwater is as shown in Table 4.10.

Simulations were then obtained day after day in unsaturated condition, by applying wetting/drying cycles with varying boundary conditions (rainwater or deicing salt)

Table 4.9 Assumed cement and silica fume compositions (weight-%)

	SiO_2	Al_2O_3	Fe_2O_3	CaO	MgO	Na_2O	K_2O	SO_3	CO_2
CEM I 52.5 PM ES	21.39	3.49	4.16	65.12	0.82	0.12	0.3	2.86	0.88
Silica fume	94.75	0.07	0.08	0.34	0.28	0.24	0.70	0.05	0.0

Table 4.10 Ionic concentration of deicing salt and rainwater ($\times 10^{-4}$ mol/l)

	pH	Na^+	Cl^-	$HSiO_3-$	CO_3^{2-}	K^+	Mg^{2+}	Ca^{2+}	SO_4^{2-}
Rain-water	7	2.83	2.82	2.5	60.0	256	10.8	20.0	1.46
Deicing salt	7	1 mol/l	1 mol/l	–	–	–	–	–	–

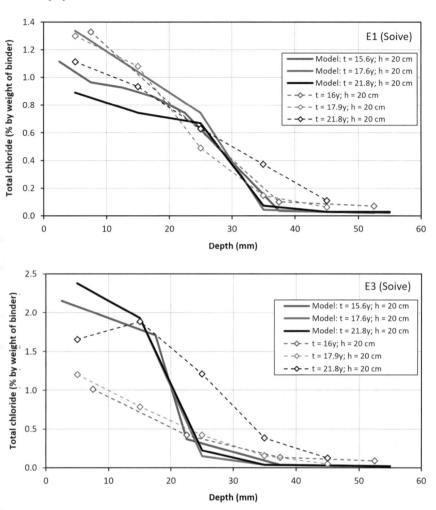

Fig. 4.8 Cerema modelling results and comparison to the measured chloride profiles. The calculations were carried out for a height from road of 20 cm only

and with temperature dependence (also obtained from the measured weather data). In the second simulation E3 (simulation with SF), only the D_{F1} diffusion coefficient was updated. The value was assumed to be equal to $8e-13$ m^2/s.

4.2.5 Double-Multi Model

The chloride ingress predictions for the Bonaduz field experiment using the Double-Multi-model were carried out by Qing-feng Liu. A description of the model can be

Table 4.11 Chloride transport properties in different phases

Field variable	Chloride ions E1/E3 (Height = 20 cm)	Chloride ions E1 and E3 (Height = 100 cm)
Diffusion coefficient in aggregates, D_A	0	0
Diffusion coefficient in bulk mortar, $D_{cem} \times 10^{-12}$ m^2/s	1.38/1.40	1.2
Diffusion coefficient in ITZs, $D_{itz} \times 10^{-12}$ m^2/s	4.14/4.20	3.6

Table 4.12 Initial boundary conditions of individual species

Field variable		Chloride ions E1/E3
Concentration boundary conditions, mol/m^3	Height = 20 cm, $x = 0$	68/100
	Height = 20 cm, $x = 100$	0
	Height = 100 cm, $x = 0$	40/33.5
	Height = 100 cm, $x = 100$	0
For 100y prediction: Semi-infinite model	$x = 1 = 200$ mm	0
Initial conditions, mol/m^3	$t = 0$	0

found in (Liu et al., 2015; Mao et al., 2019). The Double-Multi-model treats the concrete as a three-phase composite material including mortar, ITZs and aggregates with specific volume fraction (vol-70%) calculated by the parameters given in Table 4.1. For the concrete element E1 the paste volume is 25.4%, the air-volume is 4.1% and water-binder ratio is 0.46; for element E3 the paste volume is 27.7%, the air-volume is 3.4% and water-binder ratio of E1 is 0.48. Moreover, with the given annual average temperature which is ca. 8 °C (in highway), the range of chloride diffusivity in these two test cases could be calculated as shown in Table 4.11. The initial boundary conditions for chloride ion concentrations are given in Table 4.12. The effects of other ionic species are ignored in the present simulation because of their relative low concentrations in the concrete specimen and the relatively small effect of ionic interaction considered in the diffusion process (Fig. 4.9).

4.2.6 HETEK-Convection Model

The chloride ingress predictions for the Bonaduz field experiment using the HETEK-Convection model were carried out by Lars-Olof Nilsson. A description of the HETEK-Convection model can be found in Chap. 2 and in literature (Nilsson, 2000). Some model parameters were tuned using the measured chloride profiles after 1.6 and 4.3 years.

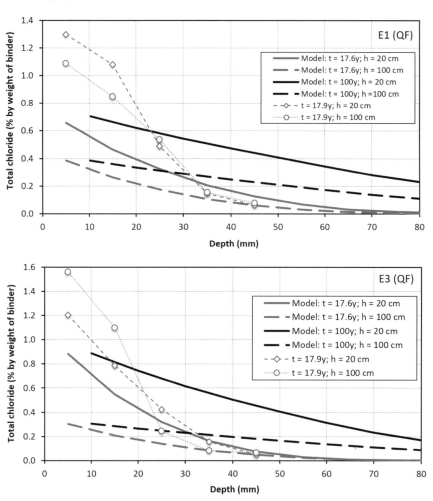

Fig. 4.9 Double-Multi-model results and comparison to measured chloride profiles after a chloride exposure time of 17.6 (17.9) years. The predictions for an exposure time of 100 years are also given. Model results are strongly dependent on height from road. The predictions for 15.6 and 21.8 years chloride exposure show similar trends and are not plotted here

The model climate was described in detail, conf. Fig. 4.10a, from the information on when deicing salt was first spread, and for how long, together with periods of wet climate and the annual variation of T and RH from weather data.

The model has 17 material parameters, for binding and transport of moisture, as vapour and liquid water, and chloride. The diffusion coefficient D_{F1} is used together with a chloride binding isotherm. None of these parameters were available for the particular concrete. Instead, an attempt was made to "calibrate" the model against short-term field data, cf. Fig. 4.5.

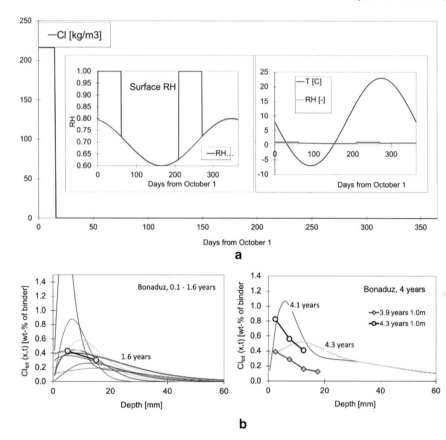

Fig. 4.10 **a** The model climate in the HETEK-Convection model is described by annual variation of temperature, RH, rainy days, days with salt. On days with salt the surface chloride is a saturated solution with a concentration of 216 **g** Cl per litre. **b** The predicted profiles during the first four years compared with the given chloride profiles after 1.6, 3.9 and 4.3 years, respectively

The diffusion coefficient D_{FI} in Fick's 1st law was set to $1 \cdot 10^{-12}$ m²/s and a chloride binding isotherm was assumed. The predictions for the first 1.6 and 4 years are compared to the given data in Fig. 4.10b.

It is obvious, from the field data and the predicted profiles, that the variations during the first couple of years are so large that such an approach is not fruitful. The real conditions in the surface region vary so much from month to month that is it decisive exactly when a chloride profile is taken compared to when salt was last spread and when the concrete surface was hit by rain. Instead, the true material properties must be determined in independent tests, especially the diffusion coefficient and the chloride binding isotherm.

Due to the lack of true material properties, long-term predictions are not expected to be correct. A comparison is shown in Fig. 4.11 between predictions made by using the assumed material parameters and the field data after 15–17 years. They do no

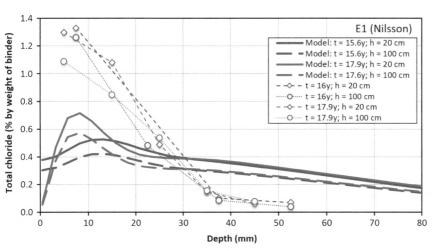

Fig. 4.11 Simulation results of the HETEK-convection model and comparison to measured chloride profiles after a chloride exposure time of 15.6 (16) and 17.6 (17.9) years for concrete element E1 only

compare at all. Most probably the decisive parameter is the diffusion coefficient D_{F1} that is assumed far too large.

4.2.7   Evaluation of Simulation Results

A way of evaluating the results of different models is to compare the 100-year predictions as shown in Fig. 4.12. Three participating models produced the chloride ingress after 100 years. The Cerema-model did not produce the 100 years results because of the required very long computation time. There is a larger scatter in results (see Fig. 4.12). Some models are height-from-road-dependent, others are not. The predicted chloride surface contents (C_s) after 100 $_{years}$ differ greatly. It seems that the magnitude of the assumed driving force of chloride ingress (i.e., assumed amount of chlorides reaching the chloride surface) differs greatly between the models. Some models assume a constant C_s and some a time-dependent C_s.

4.3   Conclusions

The model predictions for the road environment with deicing salt (i.e., Bonaduz field experiment) presented in this chapter were not carried out according to prescribed guidelines for boundary conditions and parameter fitting. This makes it hard to compare the prediction results and identify the reasons for the differences between

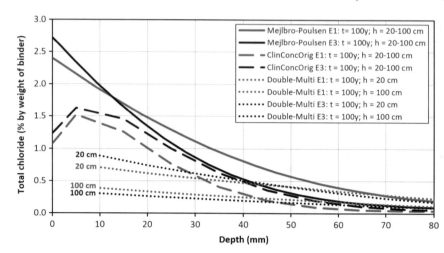

Fig. 4.12 Comparison prediction results for a chloride exposure time of 100 years for Bonaduz concrete and weather conditions

model outputs and differences between model results and field measurements. The presented simulations merely intend to illustrate the complexity of predicting chloride penetration into roadside concrete structures. The following conclusions can be drawn from the modelling attempts of the Bonaduz field results:

- Setting up a chloride ingress modelling benchmarking exercise for road environments with deicing salts is a highly complicated matter: (1) Prescribing benchmarking guidelines for differing and highly complicated models with many input parameters is a difficult task; (2) The quality of the benchmark itself, i.e., the understanding and representativity of the field experiment data, is another important issue; and (3) How to define the quality of a model simulation, i.e., what feature of modelled and measured chloride profiles needs to be compared?
- The main factor that caused the differences in model predictions and differences between predicted and measured chloride profiles could be the assumed magnitude of driving force, i.e., the amount of chlorides reaching the concrete surface and the evolution of chloride surface concentration. If different models assume different driving forces, and if the assumed driving force does not represent the actual driving force as it occurs in the field experiment, then it becomes hard to identify the quality of the modelled chloride ingress behavior.

References

Angst, U. M., Büchler, M., Schlumpf, J., & Marazzani, B. (2015). An organic corrosion-inhibiting admixture for reinforced concrete: 18 years of field experience. *Materials and Structures*, 1–12.

Bilotta, A., Di Ludovico, M., & Nigro, E. (2009). Influence of effective bond length on FRP-concrete debonding under monotonic and cyclic actions. In *Proceedings of 9th International Symposium on Fiber Reinforced Polymer Reinforcement for Concrete Structures*, Sydney, Australia, 13–15 July 2009.

Bisschop, J., Schiegg, Y., & Hunkeler, F. (2016). Modelling the corrosion initiation of reinforced concrete exposed to deicing salts. Reseach report No. 676 of the Federal roads office FEDRO (ASTRA), February 2016.

Liu, Q. F., Yang, J., Xia, J., Easterbrook, D., Lu, L. Y., & Lu, X. Y. (2015). A numerical study on chloride migration in cracked concrete using multi-component ionic transport models. *Computational Materials Science, 99*(2015), 396–416.

Mao, L. X., Hu, Z., Xia, J., Feng, G. L., Azim, I., Yang, J., & Liu, Q. F. (2019). Multi-phase modelling of electrochemical rehabilitation for ASR and chloride affected concrete composites. *Composite Structures, 207*(2019), 176–189.

Nilsson, L. O. (2000). A numerical model for combined diffusion and convection of chloride in non-saturated concrete. In *2nd International Workshop on Testing and Modelling the Chloride Ingress into Concrete*, 11–12 September 2000, Paris.

Swiss (European) standard SN EN 14629: Prüfverfahren Bestimmung des chloridgehaltes von Festbeton.

Tang, L. (2005). Guidline for practical use of methods of testing the resistance of concrete to chloride ingress. Final report of EU-funded research project 'CHLORTEST', Resistance of concrete to chloride ingress 2005.

Tang, L., & Utgenannt, P. (2007). *Chloride ingress and reinforcement corrosion in concrete under de-icing highway environment—A study after 10 years' field exposure* (p. 76). SP Report 2007.

Tang, L., et al. (2010). Validation of models and test methods for assessment of durability of concrete structures in the road environment. CBI Betonginstitutet.

Tang, L. et al. (2018). Chloride ingress and reinforcement corrosion—After 20 years' field exposure in a highway environment (p. 66). RISE Report 2018.

Chapter 5
Conclusions

Eddie Koenders, Kei-ichi Imamoto, and Anthony Soive

Abstract The comprehensive benchmark study reported in this STAR revealed various new insights in the reliability and accuracy of medium to long term model predictions for chloride ingress into the concrete cover. General conclusions are subdivided into some general benchmark observations followed by a number of conclusions drawn for the Marine submerged and the Road sprayed cases. From this a few suggestions for calibration are reported indicating particular issues that could be considered whenever developing a calibration method. Finally, an outlook and recommendations are provided addressing ideas and possibilities for future work on benchmarking chloride ingress modelling.

5.1 General

The benchmarking activities addressed in this STAR, are represented by two typical chloride ingress cases with the difference caused by the exposure conditions. The first case, indicated as *"Marine submerged"*, is based a concrete panel partly submerged in seawater, located offshore Sweden. The real data, measured from this test site was used as a starting point for the benchmark reported in Chap. 3 of this report. The second case, indicated as *"Road sprayed"*, is based on a test site along a road-side near Bonaduz in Switzerland, where chloride ingress monitoring has already been done for up to 22 years. Local conditions and details about the benchmarking has been reported in Chap. 4. These two particular cases were selected as such for this benchmark because of the differences in the surface chloride concentration with

E. Koenders (✉)
Technical University of Darmstadt, Darmstadt, Germany
e-mail: koenders@wib.tu-darmstadt.de

K. Imamoto
Tokyo University of Science, Tokyo, Japan

A. Soive
Cerema, Aix-en-Provence, France

© RILEM 2022   81
E. Koenders et al. (eds.), *Benchmarking Chloride Ingress Models on Real-life Case Studies—Marine Submerged and Road Sprayed Concrete Structures*,
RILEM State-of-the-Art Reports 37, https://doi.org/10.1007/978-3-030-96422-1_5

respect to time. Where the Marine submerged case is typically exposed to a constant surface chloride concentration and moisture conditions, the Road sprayed case has a surface free chloride concentration and hydric saturation that evolves with time. In terms of modelling, the impact of this difference is huge, since it directly impacts the ability to employ analytical formulas that presume a constant surface concentration or a saturated material. Numerical models can handle this situation explicitly, which is the reason that Chap. 4 only reports modelling results based on this class of chloride ingress models. Based on these typical differences, both benchmark cases have been used to assess the performance of the current generation of analytical and numerical models on its ability to predict the short- and long-term chloride diffusion into a concrete cover. The glance of the possible analytical and physical based models that can be employed for the prediction of chloride ingress, along with specific information on the models employed in this benchmark, is reported in Chap. 2. The chapter provides the basic information needed to understand the model backgrounds used in the benchmark. Various detailed input information employed for the two particular benchmark cases is reported in the designated Chaps. 3 and 4. From the outcomes of the benchmark cases the following conclusions could be drawn:

- Both benchmark cases revealed that reliable predictions for chloride ingress require adequate models dedicated to the problem and enabling the possibility to include the local physical and ambient boundary conditions. In particular this could be the main reason for deciding upon employing an analytical erf-function based model or a numerical model, where analytical based models have limited possibilities to include non-constant time and space dependent conditions, and numerical models do have these options.
- The Marine submerged case clearly emphasized the limitations of the erf-based function models, and its dependencies on the diffusion coefficient and surface chloride concentration. In particular the limitations on the time and space dependency of the diffusion coefficient turned out to be of particular importance whenever considering long term chloride ingress predictions. Where the time-dependency can be included via the so-called ageing factor, the space dependency cannot be considered at all, giving a clear impression about the models' limitations.
- Another clear limitation of the erf-based function models is on the fact that the model is derived for an infinite space dimension, meaning a concrete element with an unlimited thickness. In reality this is not always the case, and especially for long term predictions, a concrete structure with a finite wall thickness could be exposed to chlorides from multiple sides, which will affect the final chloride concentration inside the structural element, leading to enhanced concentrations.
- The Road sprayed case revealed that the local conditions of those concrete element being exposed to chlorides from de-icing salts strongly affect the modelling approach for chloride ingress predictions. In particular the non-sustained chloride exposure caused by the fact that the de-icing salts are only applied in winters makes it necessary to consider a non-constant surface chloride concentration in the modelling approach explicitly.

- Models that can handle non-constant surface chloride boundary conditions are mostly advanced numerical models, with an explicit formulation of this boundary condition with time. The Road spayed benchmark has shown that even with most relevant input data available an accurate prediction for the long-term chloride ingress is not trivial. Assessing the long-term exposure conditions in terms of duration, location, and the associated concentration of surface chlorides, but also temperature and moisture gradients, and the traffic intensity play a decisive role in the accuracy of the prediction. These factors, as well as many more others, may affect the surface chloride concentration of road sprayed concrete elements, which in fact represents the driving force of the long-term modelling predictions.

From the two benchmark cases reported in this STAR it has become clear that the presence of a thorough calibration method would enable a better comparison of the individual model performances and would contribute to minimize the scatter among the various chloride ingress predictions in concrete elements under various conditions. In the next section some suggestions for guidelines that could lead to a calibration method for chloride ingress models are proposed.

5.2 Guidelines for Calibration

The precision of models for prediction the chloride ingress into a concrete cover with time could be enhanced whenever considering a calibration method that would act as a common reference for all models. For such calibration method, the following general guidelines could be considered:

- Differentiate the calibration method for models which are able to consider a constant (analytical) and non-constant (numerical) surface concentration.
- Differentiate the calibration method for those cases where surface chloride concentration data is available and those cases where this is not.
- Explicitly consider the ability to take time and/or space dependency into account for the diffusion coefficient.
- Explicitly consider the time dependency of the surface chloride concentration, and consider its representativeness.
- Explicitly consider ambient effects such as moisture and thermal gradients, and other seasonal effects.
- Explicitly consider chemical/material effects such as leaching, pH change, and carbonation.

5.3 Outlook and Recommendation

A comprehensive benchmark study and analysis of the data has been elaborated and reported in this STAR. The two typical benchmark cases were selected with pre-identified critical boundary conditions with which models would be challenged. This has led to the *Marine submerged* case with a constant surface chloride concentration and the *Road sprayed* case with a variable surface chloride concentration with time. Both cases clearly conferenced a preferential use of analytical or numerical models given the specific boundary conditions. The benchmark cases also revealed significant scatters for the long-term predictions among the various models that participated. Most of the differences turned out to be caused by differences in the main input data such as the (time dependent) diffusion coefficient and the surface chloride concentration, which implicitly led to variations in the long-term extrapolations of the models. Along with this, for the Road sprayed case also the interpretation of the non-constant surface chloride concentration with time caused by de-icing salts turned out to be complicated and hard to assess, making it very difficult to finally come up with a reliable prediction of the chloride ingress with time.

From this benchmark study it became clear that a generic calibration method is paramount in order to promote the reliability of chloride ingress predictions and to reduce the scatter among the various models. A RILEM supported generic calibration method for both a Marine submerged case, and possible also for the more complicated Road sprayed case, would be strongly recommended as a logical next step in this particular field of modelling. The calibration method should prescribe the procedure on how to calibrate a chloride ingress model to a fixed set of data, provide a generic set of data for the RILEM calibration. This could be done within the framework of a next Technical Committee.

Printed in the United States
by Baker & Taylor Publisher Services